Clemens Herschel, Edward Payne North

The Science of Road Making

Clemens Herschel, Edward Payne North

The Science of Road Making

ISBN/EAN: 9783337420147

Printed in Europe, USA, Canada, Australia, Japan

Cover: Foto ©berggeist007 / pixelio.de

More available books at **www.hansebooks.com**

THE

SCIENCE OF ROAD MAKING.

BY

CLEMENS HERSCHEL, M. Am. Soc. C. E.

———

CONSTRUCTION AND MAINTENANCE OF ROADS.

EDWARD P. NORTH, M. Am. Soc. C. E.

———

NEW YORK:
ENGINEERING NEWS PUBLISHING COMPANY,
1890.

PREFACE.

The science of roadmaking was revised by Mr. CLEMENS HERSCHEL in 1877, and the paper of Mr. EDWARD P. NORTH was presented to the American Society of Civil Engineers in 1879. To the one was awarded the First Prize of the State Board of Agriculture of Massachusetts, and to the other the Norman Gold Medal of the American Society of Civil Engineers. Though considerable advance in processes and machines has been made since these dates, these two papers still contain more condensed and valuable information on a subject now attracting wide-spread attention than any similar publication of which we have knowledge. Literature on the actual detail of road-making is scarce and fragmentary; and it is with the hope that these two otherwise practically inaccessible papers may prove profitable reading to engineers and others interested in road-making that they are now reprinted.

TABLE OF CONTENTS.

PART I.

THE SCIENCE OF ROAD MAKING.

INTRODUCTION.

This treatise was written in answer to the printed circular of a Committee of the Board of Agriculture, calling for "treatises upon the science of road making, and the best methods of superintending the construction and repair of public roads in this Commonwealth."

This circular was issued about the middle of December, and as the time for writing and sending in the called-for essays was limited to January 28, the writer has thought it best, no specific character being prescribed for the treatises, to attempt to write one suitable to be so called from the stand-point of the *public*, rather than from that of the civil engineer, and, giving results rather than the methods of arriving at them, to be as concise as possible.

THE SCIENCE OF ROAD MAKING.

Starting, then, with the first of the two subjects mentioned in the circular,—the science of road making, we can divide this into three periods: 1. Laying out a road ; 2. making the road-bed, which includes all earthworks, cutting and filling, culverts, drains, bridges, even tunnels, etc.; and 3. the making of the road surface;

---

*A First Prize Treatise awarded to the Author by the State Board of Agriculture, of Massachusetts.

to which, not improperly might be added, 4. keeping the road in repair.

LAYING OUT A ROAD.

The considerations which determine the best location of a road, are those arising from the nature of the travel it is proposed to accommodate; that is, from the admissible grades, radii of curves, etc. Given two points it is desired to connect, with no intermediate point where the road is to touch, that route is the best which will cost least to build and maintain, the grades and curves being kept within bounds; and to find this location constitutes the whole problem of the engineer.

In older countries, where trade and manufactures are more settled and unchanging than in the United States, the probable future travel upon a road about to be laid out and built, forms a material element in the data that govern its alignment and grades. A very able and clever article upon this subject may be found in the Journal of the Society of Civil Engineers and Architects at Hanover," for the year 1869, and also in pamphlet form. It is in the German language, written by Launhardt, Superintendent of Highways (and a civil engineer) in the Hanoverian provinces.

The Romans built all of their roads in perfectly straight lines, up hill and down, at a very great expense, as being absolutely the shortest distance between two points. At a later period in history, it was argued that a road *must* be winding to be agreeable, and many were so built only for this reason. The modern road-builder or engineer in general, ignores any such considerations, and has for his aim only to achieve the most, at the least present and future expense.

As regards curves in roads in a hilly or mountainous district, we have then the rules never to make a smaller radius than 20 feet, and that only in extraordinary cases. On roads where long logging or other wagons may be expected, the smallest radius ought to be 50 or 60 feet; and, in general, 40–45 feet is none too much.

A rule sometimes followed in constructing mountain roads, is, where the inclination is 1 or 2 in a hundred,† heavy teams require

† In describing grades, the first figure gives the vertical height which is ascended in a horizontal distance given by the second figure. Both figures must of course be taken to refer to the same unit of length, thus: 100 feet in 120 feet, 100 inches in 120 inches, or 100 miles in 120 miles, all express the same inclination to a level plane, and are more general in their application than the ways of expressing grades in so many inches to the foot, or feet in one mile, etc., etc.

40.' and light ones 30.' radius; with a grade of 2 or 3 in a hundred, heavy teams require 65.' and light ones 50.' radius. Where a reverse curve [shaped like the letter S] occurs, there should be a straight piece connecting the two curves [Fig. 1.] On the contrary, where the two curves to be connected are concave in the same direction, the connecting link should be curved also, and not straight, [Fig. 2.] On the length of the curves the grade should be made easier than on the parts of the road immediately adjoining.

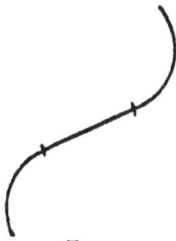

FIG. 1.

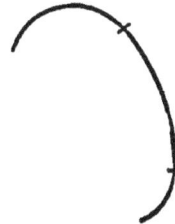

FIG. 2.

As regards grades, to start with mountain paths, we find pedestrians able to walk up an inclination of 100 in 120; mules, ponies, etc., 100 in 173. For roads, Telford's rule was, that for horses attached to ordinary vehicles to trot up a hill rising 3 in 100, was equal to walking up one of a 5 in a 100 grade.

Experiments have shown that—

1. On a road falling 2 in a hundred, vehicles would run down of themselves.

2. On the same kind of road, but having an inclination of 4 in a hundred, light vehicles had to be held back lightly, loaded ones with considerable force.

3. On a road having a fall of 5½ in a hundred, light vehicles had to be held back with considerable force, or if a brake was applied they had to be pulled, whereas heavy or loaded vehicles had to be braked to keep the horses from being speedily exhausted.

On inclinations steeper than 5 in a hundred, the rainwater running down the road is apt to do some damage to the road surface.

The regulations of different countries having a long experience in road building, such as France, Prussia, Baden, etc., vary somewhat, but the following is the general result:

In treating of roads, it often renders the subject much clearer, to divide them into three classes: first, second, and third class roads, or, as we might also say, state, county and town roads. Accepting this nomenclature, we have this: for first-class or state roads, the greatest inclination should not exceed 3–5 in a hundred; second-class or county roads, 5–7 in a hundred; third-class or town roads, 7–10 in a hundred. A road rising 10 in a hundred is not supposed ever to have any heavy teams upon it. In ascending a hill it is well and proper to decrease the grade as the top is reached, and in the same measure as the horses get tired. Thus, if a first-class road starts up hill with a grade of 4½ per hundred, it should gradually diminish to 4 and 3½ in a hundred, and end near the top with a grade of 3 in a hundred.

Launhardt, the superintendent of highways, and engineer, mentioned in the previous note, has a valuable article on the subject of the best grades for highways, in the Engineering journal there mentioned, for the year 1867; re-printed also in pamphlet form. He shows in this article that, according to the received formula that expresses the relations between the tractive force, the velocity in feet per second, and the daily working hours that go to produce the maximum amount of work that can be got out of a draught-horse, a uniform grade between any two points, except perhaps in curves, and, if desired, for resting places, is the grade that tends to enable the horse or other draught animal to produce the most work per diem.

If a grade of 4 or 5 in a hundred must needs be kept up for some distance, then it is well to have resting places 40 or 50 feet long, having a grade of only 1½ or two in a hundred, in the line of the road at proper intervals. An expedient adopted by Telford, the eminent English engineer, in order to avoid making a piece of road a mile long, on a less grade than 5 in a hundred, on account of the increased cost this would have occasioned, and yet not have this part of the road too much more tiresome for the horses than the rest, was to make the road-surface on this mile of a much better quality than on the remainder; the additional cost required for the improved road-bed amounting to only about one-half of what it would have cost to reduce the grade to say 4 in a hundred, as will be again referred to under the head of trackways. In sharp curves the grade should be only 1 or 2 in a hundred or level.

The following table gives the effects of various grades on the amount a horse can pull, and is based on calling the load a horse will pull on a level, one:—

Then, on a grade of 1 : 100, a horse can pull.......................... 0.90

"   "   1 : 50,   "   " 0.81

"   "   1 : 44,   "   " 0.75

"   "   1 : 40,   "   " 0.72

"   "   1 : 30,   "   " 0.64

"   "   1 : 26,   "   " 0.54

"   "   1 : 24,   "   " 0.50

"   "   1 : 20,   "   " 0.40

"   "   1 : 10,   "   " 0.25

To determine whether it is most advisable to go over or around a hill, all other considerations being equal, we have this rule: Call the difference between the distance around on a level and that over the hill , d, the distance around being taken as the greatest, and call h, the height of the hill.

Then in case of a first class road, we go round when d is less than 16 h.

And in case of a second class road, we go around when d is less than 10 h.

When the height of a necessary embankment gets to be more than 60 or 65 feet, a bridge or viaduct will be found cheaper, and the same measure, 60 feet, applies in case of tunnels, they being cheaper at that depth than open cuttings.

Under the head of laying out roads, something should be said of their width. Speaking only of such roads as are not apt to turn into streets from their proximity to towns and cities, it is well not to make them too broad, for the less the width, the less the cost of construction and maintenance, and a good 23 feet road is much better than a poor one 40 or more feet wide. Each rod (16½ feet) in width adds two acres per mile to the road. An agreeable form of road is to have on each, or on one side of the same, a strip 5 or 6 feet wide, sodded, and then a sidewalk equal in width to one-eighth the width of the roadway. The intervening strip above mentioned, is planted with trees and at intervals of 200–250 feet furnishes storage places, 30 or 40 long, for the materials used in the road repairs. The width of first, second and third class roadways may be given as 26, 18½ and 13 feet, with a tendency during the last ten years to have none, except in the vicinity of cities, wider than 24 feet, and the rest correspondingly narrower. In view of the changes constantly going on in this country in the value and settlement of land, it would probably be well always to *lay out* a road 50 or 60 feet wide, but to *build* the road proper of the width above indicated.

With all these rules and data in mind, the real work of actually laying out the road on the ground and on a map is next in order, and this comes so entirely within the province of the civil engineer, and is a matter requiring so much explanation and study, that it cannot well be introduced within the limits of this treatise. It is in this part of the work that a little skill and labor well spent may be productive of very great saving in the cost of the whole work and it should not be left to the inexperienced or unskilful.*

MAKING THE ROAD-BED.

Under this head are included, earthworks, drains, culverts, bridges, stay walls, etc., etc., all matters requiring a special kind of skill to construct properly. The writer believes it impracticable to write a book which shall at once be interesting to and therefore valued by the public, and of value to the professional man, and thinks an attempt so to do results always in a failure in both directions. True to the determination expressed in the introduction, he proposes, therefore, to treat under this head mainly with those parts of the subject in which the public at large is most interested, for example, the data for the cost of earthworks, general information relating to drainage, bridges, etc.

* Gillespie, in his treatise on "Roads and Railroads," gives two forcible instances of the amount those roads which might properly be called *chance* roads, can be improved by a road-maker of skill and understanding. An old road in Anglesea, England, rose and fell, between its two extremeties, twenty-four miles apart, a total perpendicular amount of 3,540 feet; while a new road, laid out by Telford between the same points, rose and fell only 2,257 feet; so that 1,283 feet of perpendicular height is now done away with, which every horse passing over the road had previously been obliged to ascend and descend with its load. The new road is besides two miles shorter. The other case is that of a plank-road built in the State of New York, between the villages of Cazenovia and Chittenango. Both these villages are situated on Chittenango Creek, the former being eight hundred feet higher than the latter. The most level common road between these villages, rose, however, more than 1,200 feet in going from Chittenango, to Cazenovia, and rises more than four hundred feet in going from Cazenovia to Chittenango in spite of this latter place being eight hundred feet lower. That is, it rises four hundred feet where there should be a continual descent. The line of the plank-road laid out by George Geddes, civil engineer, ascends only the necessary eight hundred feet in one direction, and has no ascents in the other, with two or three trifling exceptions of a few feet in all, admitted in order to save expense. The scenes of similar possible improvements are scattered all over this and the rest of the States; and these facts are still more or equally to be borne in mind in laying out new roads, where the ounce of prevention may take the place of the pound of cure.

EARTHWORKS.

The basis of all values is the daily wages of a common unskilled laborer, and in the data given below, this figure, whatever it is from time to time and in various places, must be taken as unity, or the standard measure.

The cost of earthworks may be divided into three parts—(1) cost of loosening the earth, (2) cost of transport, and (3) cost of forming the transported earth into the desired shape. The cost of the first part depends materially on the kind of earth to be handled. The cost of the second, mainly on the distance the earth is to be moved.

We find by experience, that in digging and loading or throwing 5–10 feet horizontally with a shovel, we obtain for different materials the results of the table on the next page.

TRANSPORT OF EARTH.

Throwing with a shovel.—This is to be done only from 5–12 feet in distance or from 5–6 feet vertically. To throw 5 feet vertically, costs as much as 12 feet horizontally, that is to say, if 30 feet horizontally cost per cubic yard, one day's wages divided by 8.4 the same distance vertically will cost about 2½ times as much, or more exactly, one day's wages divided by 3.5, whence is seen the economy of using windlasses, etc., instead of " stages,"† in shovelling earth vertically. The table gives the cost of shovelling earth certain distances, expressed in the number of cubic yards a laborer's day's wages will pay for.

DISTANCE OF THROW IN FEET.	Vertical or Horizontal.	Whether done at one operation, or by means of so-called "stages."	Number of cu. yds. which can be transported at the cost of one laborer's day's wages.	Remarks.
0–10, . .	Horizontally,	No "stages."	23.5	
10–20, . .	"	1 stage.	12.6	} Wheelbarrow
20–30, . .	"	2 stages.	8.4	} cheaper.
0–5, . .	Vertically,	No stages.	14.1	
5–10, . .	"	1 stage.	8.8	

† By a "stage" is meant the operation of one shoveller lifting and throwing what another has thrown in front of him.

Number.	KINDS OF EARTH.	In Parts of a Laborer's Day's Wages			Amount to be added for keeping and repairing tools.
		Cost per cubic yard to loosen.	Cost per cubic yard to load in wheelbarrows.	Cost per cubic yard to load in carts or wagons.	
1	Loose earths, which are loam, sand, etc., inclusive of loading......	$\frac{1}{7}-\frac{1}{9}$	—	—	$\frac{1}{20}$
2	Heavier earths, such as sticky clay, which does not readily leave the shovel, etc........	$\frac{1}{6}-\frac{1}{7}$	$\frac{1}{16}$	$\frac{1}{12}$	$\frac{1}{20}$
3	Earths which must be loosened with a pick before they may be shovelled	$\frac{1}{4.5}-\frac{1}{6}$	$\frac{1}{16}$	$\frac{1}{12}$	$\frac{1}{16}$
4	Solid banks of gravel or clay, earths containing boulders, etc., in which one man only loosens as much as another man shovels..	$\frac{1}{4}-\frac{1}{4.5}$	$\frac{1}{16}$	$\frac{1}{12}$	$\frac{1}{12}$
5	Same material, worst kind, brick and mortar heaps, earth full of roots, etc., in which it takes two men to loosen what one man shovels	$\frac{1}{3}-\frac{1}{3.5}$	$\frac{1}{12}$	$\frac{1}{8}$	$\frac{1}{16}$
6	To break up stone which is in layers or seams, requiring the use of the crowbar only, but no blasting.......	$\frac{1}{2.5}-\frac{1}{3}$	$\frac{1}{12}$	$\frac{1}{8}$	$\frac{1}{16}$
7	Blasting rocks in an open cut, according to the hardness of the rock, to the position of the seams, etc.*	$\frac{1}{2.6}\ \frac{1}{1.5}\ \frac{1}{1.25}\ 0.75$	$\frac{1}{12}$	$\frac{1}{8}$	$\frac{1}{16}$
8	In forming and shaping embankments	$\frac{1}{2.5}$	—	—	$\frac{1}{30}$

*To excavate rock to a given line and level, that is, to trim a cutting may cost double these figures per yard.

Wheelbarrows.—The usual distance of transport suitable for the use of wheelbarrows is 100-200 feet. In exceptional cases it may be more, but perhaps never above 500 feet and then only for moderate quantities. In going up hill, the greatest inclination is to be not more than 1 in 10, and a man can push only ⅔ as much on this inclination as on a level. 3 feet vertical transport costs as much as 90-100' horizontally. Whenever possible, planks should be laid for the wheel-barrows to run on. The best timber for this purpose is beechwood and the cost of keeping such planks is only about $\frac{1}{40}$ or $\frac{1}{60}$ per. cent. of the cost of transport per cubic yard.

Distance of Transport in Feet.	Number of trips per day of ten hours, made with one man at barrow, and one to load.	Contents of wheel-barrow load in cubic feet.	Number of cubic yds. which can be transported at the cost of one laborer's day's wages.
10–20, ...	120	2½	23.5
20–50, ...	110	2½	16.9
50–70, ...	100	2½	14.4
70–100, ...	98	2½	13.8
100–150, ...	96	2½	13.3
150–200, ...	94	2½	12.8
200–250, ...	92	2½	12.4
250–300, ...	90	2½	12.0
300–350, ...	88	2½	11.6
350–400, ...	86	2½	11.2
400–450, ...	84	2½	10.9
450–500, ...	82	2½	10.5
500–550, ...	80	2½	10.2

PATENT PORTABLE RAILROAD AND HAND-CARS.

These have lately been introduced in this country, and appear to be coming into general use and favor. The company owning this improvement, as it seems to have a right to be called, claim, that by means of their track and cars, which can be used everywhere that a wheel-barrow or a horse-cart can go, and in a great many places where these vehicles cannot go, they affect a very large saving, as much in some cases as ⅘ of the cost by the other means of transport. There are no data published as yet to make tables from, similar to the foregoing; from the company's pamphlet, however, one given case which occured on Staten Island in 1867, may be analyzed and tabulated as follows:—

Distance of transport, in feet,   .   .   .   ·   .   550
Number of trips per day of ten hours, with one man at
  two cars, and two to load,   .   ·   .   ·   ·   150
Contents of car in cubic feet,   .   ·   .   ·   .   11.34
Number of cubic yards which can be transported at the
  cost of one laborer's day's wages,   .   ·   .   ·   60

ONE-HORSE CARTS.

The table for this kind of transport may be stated about as follows.   1 foot vertical costs as much as 14 horizontal.

Distance of Transport in feet.	No. of trips made per day of ten hours, assuming only four minutes to load, dump, &c. per trip.	Contents of cart load in cubic feet.	Number of cubic yds. which can be transported at the cost of a laborer's day's wages.
300, ...	86	8	17.1
500, ...	67	8	13.6
1,000, ...	43	8	8.6
1,500, ...	31	8	6.4
2,000, ...	25	8	5.0
2,500, ...	21	8	4.3
3,000, ...	18	8	3.6

Ox-cart transport is 10 or 12 per cent. cheaper than the above, but takes more time.

Other methods of transport, such as horses or engines on temporary tracks, would hardly ever be applied to road-building, but belong more properly under the head of railroad construction.

SHRINKAGE.

In calculating the cost of earthworks, the so-called shrinkage of earth must not be overlooked.   Earth occupies on the average $\frac{1}{10}$ less space in embankment than it did in its natural state, 100 cubic yards, shrinking into 90.   Rock on the contrary, occupies more space when broken, its bulk increasing by about one-half.

The shrinkage of gravelly earth and sand may be taken at 8, of clay 10, loam 12, surface soil 15, and of "puddled" clay 25 per cent. The increase of bulk of rock is 40 to 60 per cent.

To make use of all these data in calculating the probable cost of a piece of road, there are of course still wanting the equally essential factors which give the number of cubic yards to be dug and moved and the distance of transport. These are got from the plan, profile and cross-sections of the proposed work, an engineer's knowledge being requisite to make the necessary drawings and calculations.

DRAINS AND CULVERTS.

The drainage of roads is of two kinds, surface and sub-drainage. The first provides for a speedy removal of the rain-fall on the surface of the road and the cutting and embankments on which it is carried; the second, for the removal of that part of the rain-fall which nevertheless does penetrate into the body of the road-covering. With a perfect sub-drainage the winter's frost, having no water to act upon within the body of the road, is robbed of its great power to destroy the same, and it also prevents the road-surface from becoming soaked and thence destroyed in the summer. The need of surface drainage is self-evident. This last named is to be provided for at this stage of the building of the road, the sub-drainage being more properly a part of the building of the road-covering or top. For this purpose ditches, one on each side generally, are absolutely necessary, both when the road is on a level with the surrounding country and when it is in a cutting. They may become necessary also in the case of embankments: for example, when an embankment is built across wet ground. Where these side ditches cross under the embankment we have a culvert: also whenever any small valley, having a constant or intermittent stream of water, is crossed by such an embankment. It is very bad policy to make such culverts of wood, unless indeed they are so situated as to be constantly under water; the cost of replacing them after the embankment and road has been built over them is disproportionately great. They should be made of stone, or brick; lately vitrified stone-ware, or cement drain-pipe, oval or egg-shaped, has been used to advantage in their construction.

All ditches, drains and culverts should have a fall throughout their entire length. Their size will depend on the amount of water they may be expected to carry, and this again on the rain-fall that may occur on the area which they drain. Extraordinary showers have occurred of 2 inches in half an hour but only over a very limited area, and 2 inches in an hour may be taken as a large allowance. This is the basis of the Central Park drainage calculations, and is larger than usually taken, none too large however for safety.

The determination of the proper width and height of culverts, that will enable them to pass the requisite quantity of water without damming it up, is a question in practical hydraulics, easily enough settled, in cases of doubt, by the proper gaugings and observations made upon the spot, but which is answered only in a very crude and imperfect manner by any general rules that may be given. And yet it may prove a very important question at times. There is now (1877) pending in Massachusetts, a suit for damages, that may involve claims to the amount of rising half a million dollars, in which one great centre of attraction is nothing but a simple railroad culvert, and the question: Was it as large as it ought to have been? and the writer passes every day, when at home, by a culvert, which for some 150 or 200 years has dammed the waters of a brook back about 3 miles, from 1 ft. to say 20 in. at the culvert vertically, and done this right along two or three times per annum, and at the present time it contributes in this manner, more than its proper share towards the flooding of about 500 cellars. These two cases may serve to call attention to the great damage that may accrue from making culverts too small, and to show whence comes the rule: in cases of doubt, make the culvert plenty large enough. The following rough and approximate rules for determining the quantity of water that a culvert will be called upon to pass through it, are taken from a German pocket-book for road engineers. Compute the cubic feet per second from the drainage area that lies above the culvert, and, for the different lengths of valley from the corresponding rain-falls per hour. (The rain-fall is given in inches per hour, instead of in decimals of a foot per second, only for the purpose of avoiding the printing of long decimals).

LENGTH OF VALLEY IN MILES.	INCHES PER HOUR.
2.5 or less,	1.2
2.5 to 5.	0.75
5. " 7.5	0.45
7.5 "10.	0.30
10. or more,	0.15

As culverts grow larger and wider with the amount of water they are to pass under the road, they develope finally into

BRIDGES.

Bridge-building is a life's study, taken by itself, and in some of its parts it is not half appreciated and known as yet among the public. Prominent among these is beauty of design and *appropriateness to the situation.* There is perhaps nothing else that will so much improve the appearance and attractiveness of a road as a beautiful bridge. So also in cities we find that a street will of its own accord, seemingly, improve in appearance, when a good and handsome bridge has been erected on its line, the owners and builders of the adjoining buildings taking the bridge for their pattern and model. Nor must it be supposed that a handsome bridge must necessarily cost more than an inappropriate or homely, uncouth structure; it need never be the case. Very often the chief beauty of a structure lies in the fact of its carrying the most with the least expenditure of material. No one bridge is proper in every situation, and herein many mistakes are made. The correct way to build a good bridge, is the same or a similar way to that followed in first-class buildings, namely, to have plans drawn for the same and receive estimates and offers to build according to these plans. It is not well to allow the offices of designer, superintendent and contractor to be united in one person or firm, and is expecting too much from human nature.

MAKING THE ROAD SURFACE.

There are two subordinate kinds of road surface, if the term road can properly be applied to them, namely, that of foot and riding paths; these may be disposed of first, before proceeding to the more important consideration of the road surfaces proper, those used by vehicles of all descriptions.

Footpaths.—For the surface of a foot path little solidity is necessary, except in city sidewalks, which are not supposed to be treated of here, but we do need a material that shall become and stay compact soon after it is laid. Coarse sand, screened gravel, stone chips and dust, make good paths; should these materials be too free from any earth or clay, a little of the same may often be added to advantage to act as a binding material. Wherever the ground underneath the surfacing is not porous or likely to remain porous enough to let all the water that may soak through drain away, a layer of such porous material must be filled up before the top surface is put on. Oyster shells, or large stone chips, gravel stones or pebbles, etc., make a good foundation of this sort. The top covering should have a slope, best in both directions from the centre of the path towards each side of about 1 in 16; the thickness of the foundation course to be 3 to 5, and that of the top 3 to 4 inches.

No *gravel* path, or side-walk, will afford good walking at the season of the year when the frost is coming out of the ground. Carting on more gravel is in vain; it is often no better than mere foolishness. If village communities will get this idea firmly into their minds, and, instead of a fruitless struggle against the laws of nature and of gravel, will build stone screening sidewalks *with a good foundation course* underneath, as above described, or else some sort of hard sidewalk covering, they will save themselves much expense, many muddy feet, and no small amount of consequent and annual discontent, not to say profanity and ill feeling.

Heavy rolling and wetting down will save much time in finishing the whole process; the roller should be used unsparingly and throughout the whole construction of the path, on the foundation, as well as on the top.

Riding-Paths.—From the nature of the travel these are intended to acommodate, their surface must be of a peculiar kind. Inasmuch as a horse, in galloping, tends to throw the soil he treads on backwards with his hoofs, the surface must be kept somewhat loose and soft to make riding on it easy and agreeable.

This requirement makes it impossible to have any slope on the surface (the loose material would wash away if there were any), and hence we must rely here wholly on sub-drainage, and not attempt any surface drainage. The top is made of coarse sand, *free from clay* or other binding material, laid on two and one-half to three and one-half inches thick, and spread out level. Under this is a solid foundation, about four inches thick, made of coarse gravel and clay, and having a slope of about 1:20, so that the water will run off along its top surface to either side, where it must further be disposed of by drains or ditches. In case of riding paths too wide to be so simply built, the sketch shows the method to be used. The

foundation is made in several slopes, at the lowest parts of which are placed drains, running in the direction of the path, but communicating from time to time with the sides of ditches or drains. Should, however, the ground underneath be porous enough, the drains may be dispensed with; and if in their stead holes be dug along the lowest lines, marked *a, a*, and these filled with large stone, the water will, through them, drain away into the ground.

Roads.—To make a good road surface is a very simple operation after it is only once understood, and, the fundamental principles thereof once comprehended, they can hardly be forgotten. Everything connected with the construction, the use and maintenance of roads, was, in times past, before the invention of railways, the subject of exact observations and experiments, many and varied in character.

Old engineering works that treat of road-making are on this account excellent reading upon this subject at the present day. Upon road *construction* no less than upon the need of better road-legislation. Some, perhaps the most of the evils we suffer, in the shape of bad common roads, are merely the result, the necessary consequence of our bad systems of common road management, which are derived from our antiquated legislation upon that subject. Legislation of this kind has changed but little in a hundred years, and is producing the same evils to-day, that it did a hundred years ago. Hence it is explainable, that the complaints concerning bad roads, and bad road management which we read in books of fifty and of sixty years ago, sound to our astonished ears as though they had been written but yesterday. On this subject may be consulted: The life of Telford, the great English road builder, who died some fifty years ago, (also among " The Lives of the Engineers," by Samuel Smiles), "A treatise on Roads," by Sir H. Parnell, 1833, and other works of former date.

Besides this, we have the results of a great number of years of experience in older countries, and there would seem to be little to invent, but much to learn, in this branch of construction. Though less progressive than other branches, there are nevertheless improvements in road-making, especially in road-making machinery and tools; and no treatise on this or any other living subject can be considered complete in a very few years after it is written.

Ancient roads were made with a surface as nearly resembling the solid rock as possible. So, in China, roads were made of huge granite-blocks laid on immovable foundations. In time these became worn with ruts, especially in the joints or seams of the stones, and the surface generally so smooth that animals could hardly stand, far less trot on it. They are now for the most part deserted, and left to be covered by land-slides, etc., to one side of the new roads of travel.

The invention of McAdam consisted in having no large stone at all on the roadway, but having it all pounded into fragments and spread over the road-bed. This has, without fear of efficient contradiction or shadow of doubt, been proved by trial to be a worthless proceeding, though at one time popular, and even now too often done, either from ignorance or laziness. The separate fragments of stone, have no bond among themselves, are liable to sink into the underlying ground or road-bed, evenly or unevenly as may chance, more in one place than another, and thus never come to rest or to an even top surface. Between these two extremes of an ancient Chinese solid rock road and that of McAdam, lies the true principle of road-making, which consists in giving every road two component parts; one,—the foundation,—to be solid, unyielding, porous, and of large material; the other—the top surface—to be made up of lighter material, and to be made to bind compactly and evenly over the rough foundation. This constitutes the whole principle to be followed; and let it be repeated, that to dump the road material directly on the ground, without first preparing a foundation for it, as is so frequently done, is a waste of time, labor and materials, by no possibility resulting in a good road. On this one fundamental idea, which is never abandoned, however, there are a number of variations. Besides these roads, whose characteristic is the foundation they are all built on, we have paved roads, or pavements, of a great many kinds, and roads with trackways, also of various kinds.

FOUNDATION ROADS.

The roads of this kind, with macadam for the top surface, are called Telford roads by English writers, from Telford, who first built them in England. The Central Park " gravel roads " belong under this head, gravel taking the place of the macadam of the Telford roads. These foundation roads are of far greater importance than any other kind for State, county or town roads, also for parks and driveways. The top surface of all these roads must have a certain inclination, to cause efficient surface drainage. Various authorities give various rules for the amount of this inclination or side-slope. It would seem just that it should depend on the nature of the top covering, being less for more solid than for looser or softer materials, and also on the grade of the road.

In Baden, one of the smaller German States, but which is worthy to be taken as a model in matters of road-building, and in France, the rise at the centre is given as $\frac{1}{40}$-$\frac{1}{60}$ of the width of the road, according to the nature of the material; that is, inclinations of 1 in 20, and 1 in 30. The rules in Prussia prescribe inclinations of 1 in 24 for roads falling more than 4 in a hundred; 1 in 18 for roads on a grade of between 2 and 4 in a hundred; and 1 in 12 for those on a grade of less than 2 in a hundred. When first built, the centre should be made some four inches too high, to allow for after settling.

Macadam Top. — The cross-section of such a road is about as drawn;  the thickness of the foundation $b=a$, the thickness of the top covering at the centre, and is six, four or five and three and one-half inches in thickness for first, second and third class roads. If the stone for the foundation—for which most anything will do, and that kind should be taken which is cheapest to procure—happens to be got out cheapest in larger pieces than the above dimensions, it will do no harm. This foundation course is sometimes set so as to present an inclination on top, and the cover then put on of a uniform thickness over the whole breadth. This is perhaps best, but is somewhat more expensive. It will do, in nearly all cases, to set the foundation course on a level, or as near so as the stones will allow, and then make the top crowning, by making the covering say three-quarters of an inch or an inch less thick at the edges than in the centre. The stones forming the foundation should not be set in rows, nor ever laid on their flat sides, but set up on edge and made to break joints as much as possible; that is, set up irregularly. After they are set up, the points that project above the general level may be broken off, and the interstices generally filled with small stone. More or less care and work are necessary in this part of the operation, according to the importance of the road and the depth and character of the material used for the top covering. To roll the road at this stage is to be recommended; afterwards it becomes a requisite. The point never to be lost sight of, is that this foundation course must remain porous, must be *pervious* to water, so that all rain-water that shall soak through the top covering will find, through it, means of escape to the ground underneath; thence, according to the nature of the subsoil, it is left either to soak into the ground, or must be further led away by appropriate drains.

Of very great importance is the *material* used for the top or road covering.   In the order of their value for macadam, we have.

    I. Basalt.

   II. Syenite and Granite.

   III. Limestones.

   IV. Sandstones.

It will be evident, that a much greater quantity of the soft stones would be required to repair a certain road, than of a harder kind, and on a road lying out of the way of a hard stone quarry or deposit, the question will arise which is cheapest, to pay more for the raw material and get good stock, or pay less and use the worse?   There have been some interesting results in places where this matter has been the subject of experiment, continued for a number of years.   Thus, on a road in Baden which was formerly macadamized with rock costing only fifty cents per cubic yard, it

FIG. 5.   PERSPECTIVE VIEW OF NEW PATTERN CRUSHER.

was finally found cheaper, to take harder rock from a distance costing one dollar and seventy-eight cents per cubic yard, the saving being both in less *quantity* of material used and less *labor* required in repairs.   Just where the limit is, must be found in each case by long continued experiment, which is well worth the trouble to make, both to save expense and also to have the best possible road, the harder material making a road better at all times, at the same or less cost. After the right kind has been determined, none other should be mixed with it, and should any inferior piece accidentally or designedly get into the stock to be broken up, it should be picked out and thrown aside. The stone is broken up into macadam, either by hand or machinery. Wherever any considerable quantity of macadam is in present or future demand, a stone-breaker is certainly a saving over hand-labor, though it is difficult to draw a line exactly, where hand-labor or machine labor is cheapest. Probably no town that pretends to keep thirty or forty miles of road in good repair, ought to be without one of these labor-saving machines. Those most in use are made by the Blake Crusher Co., of New Haven, Conn., and the following is taken from their circular.

Their machine has been patented in the United States and in several foreign countries, and is now in use in almost every country on the globe. It is simple and compact, and being complete in itself, requires no extraneous support or fixtures. Two patterns of the machine are now sold: the old, or " Lever Pattern," and the new, or " Eccentric Pattern." The figures, and following description, refer to the last named machine. Fig. 5, is a perspective view of the machine, entire. The frame which receives and supports all the other parts, is cast in one piece, with feet to stand on the floor or on timbers. These feet are provided with holes for bolts, by which it may be fastened down if desired; but this is unnecessary, as its own weight gives it all the stability it requires. The flywheels are on a shaft which has its bearings on the frame, and which between these bearings, is formed into a short crank. On the same shaft is a pulley, to receive a belt from a steam-engine or other driver.

Figure 6 shows a side view or elevation of the parts in the machine in place as they are presented to view through the side of the frame. The circle D, is a section of the fly-wheel shaft, which should make from 225 to 250 revolutions per minute. The circle around D is a section of the eccentric. F is a pit-

man or connecting **rod,** which connects the eccentric with the toggles G, G, which have their bearings, forming an elbow or toggle
joint. H, is the fixed jaw; this is bedded in zinc against the
ends of the frame ¼ inch thick. P, P, are chilled plates against
which the stone is crushed; when worn at the lower end they can
be inverted, and thus present a new wearing surface. The cheeks
I, I, fit in recesses on each side, and hold the plates in place; by
changing the position of the cheeks from right to left, when worn,
both will have a new surface. J, is the movable jaw; this is supported round the bar of iron K, which passes freely through it, and
forms the pivot upon which it vibrates. L, is a spring of India
rubber, which is compressed by the forward movement of the jaw,

FIG. 6. SECTIONAL VIEW OF NEW PATTERN CRUSHER, WITH PARTS LETTERED
FOR CONVENIENCE IN DESIGNATING PIECES WANTED FOR REPAIRS.

and aids its return. M, M, are bolt holes. B, is the fly wheel. C,
is the driving pulley. Every revolution of the crank causes the
lower end of the movable jaw to advance towards the fixed jaw
about one-fourth of an inch and return. Hence, if a stone be
dropped in between the convergent faces of the jaws, it will be
broken by the next succeeding bite; the resulting fragments will
then fall lower down and be broken again, and so on until they
are made small enough to pass out at the bottom. The readi-
ness with which the hardest stones yield at once to the influence of
this gentle and quiet movement, and melt down into small **frag-**

ments, surprises and astonishes every one who witnesses the operation of the machine.

It will be seen that the distance between the jaws at the bottom
limits the size of the fragments This distance, and consequently
the size of the fragments, may be regulated at pleasure. A varia-
tion to the extent of ⅝ths of an inch may be made by turning the
screw-nut W, which raises or lowers the wedge N, and moves the
toggle-block O forward or back. Further variations may be made
by substituting for the toggles G, G, or either of them, others that
are longer or shorter; extra toggles of different lengths being furnished for this purpose.

This machine may be made of any size. The builders have patterns for some 13 different sizes on hand at the present day. Each
size will break any stone, one end of which can be entered into the
opening between the jaws at the top. The size of the machine is
designed by the size of this opening; thus, if the width of the jaws
be ten inches, and the distance between them at the top five inches,
we call the size 10×5. The following table shows the principle
facts that relate to the sizes of machines that are used, generally,
for the making of road-metal.

No.	SIZE, or receiving capacity.	Product per hour in cubic yards. (See Note.)	Weight of Heaviest Piece.	Total Weight.	Extreme Dimensions.			Driving Pulley.		Proper Speed.	Horse power required.	Cash Price Delivered on board May, 1877.
					Length	Bre'dth	Height.	Diam.	Face.			
	Inches.				ft. in.	ft. in.	ft. in.	ft. in.	inch.			
2	10×7	Five.	4339	8000	5 3½	3 8	4 5	2 0	7½	250	6	$900
*3	15×5	Six.	4700	9100	8 7	5 0	5 0	2 4	8	180	9	1035
*4	15×7	Six.	5890	10490	8 7	5 0	5 0	2 4	9	180	9	1125
5	15×9	Seven.	6436	13360	6 5	5 0	5 11	2 6	9	250	9	1234

NOTE.—The amount of *product* depends on the distance the jaws are set apart, and the
speed. The product given in the Table is due when the jaws are set 1 1-2 inches open at thy
bottom, and the machine is run at its *proper speed* and *diligently fed*. But it will also vary
somewhat with the character of the stone. Hard stone or ore that breaks with a snap will go
through faster than Sand stone.

To make good road metal from hard compact stone, the jaws
should be set from 1¼ to 1½ inches apart at the bottom. For
softer and for granular stones they may be set wider.

A cubic yard of stone is about one and one-third tons.

In getting an engine to drive one of these Crushers, it is advisable to have one of greater power than just what is stated in the table as required. It is much more economical to use 9-horse power from a 12 or 15 horse, than from a nine or ten horse engine. The machine may be driven by any power less than that given in the table, yielding a product per hour smaller in the same proportion.

10 x 7—(No. 2.) Will take in a stone 10 inches wide and 7 inches thick, and is quite an effective machine. It may be set to break to any size down to ¾ inch, and can be used for the same purposes as the 10 x 5, but receiving pieces two inches thicker is preferred in many cases. It will do a good deal in the preparing of road metal. It is one of the most salable sizes.

15 x 5 — (No. 3). This machine takes in a stone only 5 inches thick, but being 15 inches wide is a more effective machine than the 10 x 5, but is used for same purposes when a larger product is required.

15 x 7 and 15 x 9—(Nos. 4 and 5). These are the sizes most salable, and best adapted to general purposes. They are the sizes almost uniformly used for breaking stone for McAdam roads and Ballasting railroads and for concrete. They are also used extensively at smelting furnaces — also at copper and other mines, to take the product of the coarse breakers and reduce it to proper size for feeding under the stamps.

When broken by hand and for country roads, the stones should be broken on the storage places already mentioned, which are to be established along the side of the road every 200 to 250 feet. The laborer is not to pound the stones on a heap of such, but to use one large stone as a sort of anvil to break the others on. He is to use a light hammer, except for pieces containing more than four or five cubic feet, and may use a ring with a handle attached to hold the stone he desires to break.

In order that the road shall get an even surface, the macadam must all be of one size, and the proper size for the macadam depends on the degree of hardness of the rock. If too small, it turns to dust; if too large, the top will not pack even. The size is regulated by the use of a ring as a gauge, — every stone being obliged to be capable of falling through this ring in any direction it may be dropped. Hard stones should be one to one and a quarter, softer ones one and a half, and the softest two inches in diameter. Larger sizes give less perfect roads. In loading and otherwise handling macadam, a many and close-pronged pitchfork should be used instead of a shovel, so as not to mix in any earth or sand, and so sift out the stone dust and chips.

The macadam being properly prepared and loaded up, it is spread over the foundation in two or three successive layers. Each layer *should* be rolled, but the top and last one *must* be rolled to make a good road. Nor will rolling alone do the work. Two other helps are needed: the use of a building material, to act as a cement between the broken stone, and sprinkling. It is diffi-cult to prescribe in words just what to use as binding material, and just how much to sprinkle and roll; common sense will in most cases be a safe enough guide. In the macadamized streets of Paris the rule is to roll till a single piece of macadam placed under the roller, will be crushed, without being pressed into the road surface. Gravel somewhat mixed with clay by nature, but not too much, is probably best as a building material. Clean coarse sand is very good. Other substances will do, where it would cost too much to procure either of the above.

A late writer in the "Journal of the Society of Civil Engineers and Architects," at Hannover, calls attention to the practice in Bo-hemia of making foundation roads by setting first the foundation course, spreading a thin layer of the binding material on that, and the broken stone on top of this again. The subsequent rolling has the effect of forcing the binding material, slowly and gradually, from beneath upwards into and through the broken stone. The writer states that he himself has tried a system of road construction that consists of a combination of the two methods hitherto used, with good results, namely: first, a foundation course, a thin layer of binding material on this, then the broken stone, another thin layer of binding material, and then wet down and roll.

The subject of rollers is one demanding some attention. In general, people are apt to over-estimate the value of a roller with respect to its weight. It will be evident on reflection that a roller should be as heavy per inch in length of roller, as a loaded wagon wheel is per inch of tire; or, in other words, if we have a wagon

with tires two and one-half inches wide and on each wheel a load
of say one ton, the roller should weigh two-fifths ton for every
inch in length, or a roller three feet long should weigh about fourteen and one-half tons, or else a wagon as above described would
exercise more pressure on the road-bed per square inch than the
roller, and consequently would cut into the rolled surface and produce ruts.

The proper width of tire, or proper load upon any vehicle, for
a given width of tire, is a question that occasionally attracts attention. Bokelberg, a good German authority on the subject, in an
article in the "Journal of the Society of Civil Engineers and
Architects," at Hannover, 1858, comes to the conclusion that for
four-wheeled vehicles, upon a broken stone road, the loads should
vary with the widths of tires, as follows:

WIDTH OF TIRES. INCHES.	LOAD IN LBS.
2 to 3	5,000 to 6,600
3 " 4	6,600 " 8,800
4 " 5	11,000
5 " 6	13 000
6 " 7	15,000
7 and over.	16,500

Further conclusions are: that the best width of tire, measured
when they are new, for the transportation of freight, is from four
to seven inches; this width being best for the easy traction of the
load no less than for a minimum wear of the road surface. To
make the tires wider than seven inches does not diminish the force
required to move the load, and unnecessarily increases the dead
weight of the wagons.

Road-rollers are of two principal kinds: those pulled by horses
and those propelled by steam. The latter are for many reasons
the best. In the first place they can be made as heavy as desired,
without proportionally increasing the cost of propelling them, and
being self-propelling, the only track they make is that of the roller,
whereas with horse rollers, the hoof-marks of the horses are a
great objection. Then again in the amount of work they will do
at a certain cost, they excel horse rollers. They may be briefly
described as a sort of locomotive mounted on three or four very
broad and heavy wheels, these latter being the road rollers.

An excellent pamphlet on the subject of steam road-rollers is
the "Report on the Economy of Maintenance and Horse-draught
through Steam Road-rolling," by Frederick A. Paget, E. & N. F.
Spon., 1870. Readable articles on the same subject are: "Steam
Road-roller," Engineering, Oct. 4, 1867. "Paris Kind of Steam
Road-roller," Engineering, May 7, 1869   "Cost of Operating

Steam Road-rollers," Engineering, June 18, 1869. "Good Steam Road-roller," Engineering, Jan. 14, 1870. "Economy of Steam Road-rolling, Engineer, April 1, 1870. "How to use the Road-roller during alternate Thawing and Freezing," Annales des P. & C., 1877, p. 125. In the spring and fall on finished roads, and occasionally during the first construction or reconstruction of roads, the surface becomes *sticky* mud, and to roll the road at those times, or to travel on it, tears up the covering and spoils the whole. If at such times the *roller be constantly sprinkled and kept wet* while it is being used, it will shed the mud or road covering, instead of tearing it up, and will consolidate the road in a very superior manner. And this method requires less water (only about one gallon to one and a half gallons per 100 feet of travel) than the method formerly used under these circumstances of converting the *sticky* mud into liquid mud, by copiously wetting down the whole road.

There are several varieties in use in France and England, and two at least of the English kind have been imported into this country, one for the New York Central park, the other for the Arsenal Grounds in Philadelphia. The cost of the Central Park steam road-roller made by Aveling & Porter, of Rochester, Kent, England, was about $5,000, set up in New York, and the amount of work it will do in one day at a running expense of $10, has been given as equal to that of a seven-ton, eight-horse road-roller in two days at $20 per day, or, in other words, it will do the same work at one quarter the running cost and in one-half the time, of a first-class horse road-roller.

Since 1870 many other steam road-rollers have been bought by various parties in the United States. Thus there is one owned by Daniel Brennan, a road contractor, in Orange, N. J.; the city of New Haven, Conn., has run one with great success for several years; after many years of agitation on the subject, the city of Boston now owns and operates a steam road-roller; and so on.

The best horse road-roller of which the writer has any cognizance is the one shown by the annexed drawings in plan, elevation, and in perspective.

(The town of Malden, Mass., has built a horse road-roller, according to the plan here described.)

It originates in Chemnitz, Germany, but can, of course be easily made by any machine-shop or foundry. The hollow roller is made of cast-iron, and is so arranged that it may be filled with water when it is to be used in heavy rolling; when not in use and about to be moved from place to place, the water is allowed to run

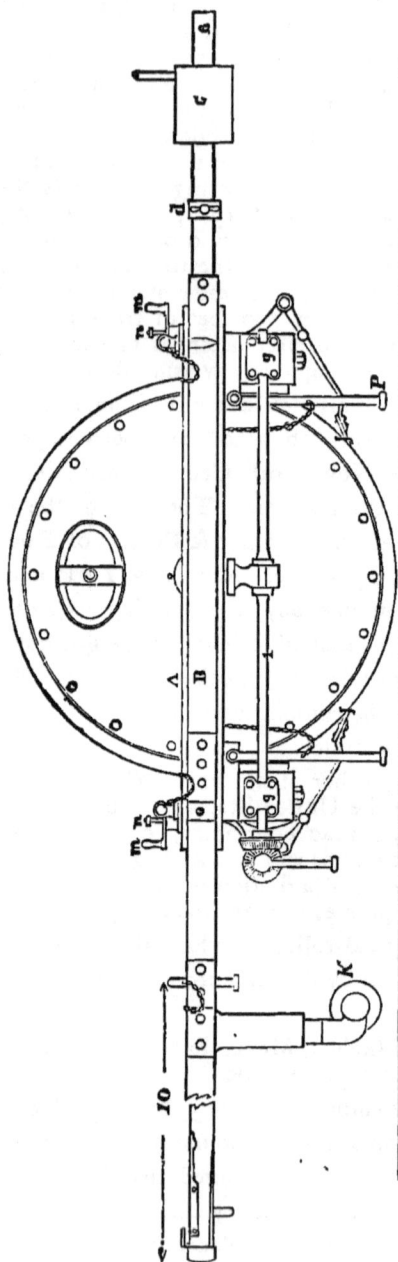

Circular Frame Horse Road Roller—Perspective View.

out, thus materially lessening the load. A circular cast-iron frame A, surrounds the roller, and carries the axle bearings of the same. The outside of this frame is turned to form a groove in which a strong wrought-iron ring is fitted in such a manner that it will turn easily around the former. This wrought-iron ring consists of two semi-circular parts, at whose junction the pole is attached on one side, and on the other an extension bar, carrying the balance weight *c*, which may be shifted by means of the set clamp *d*, or turned up by means of the hinge *b*. Pins going through the holes at *e*, fasten this ring or allow it to be turned for the purpose of pulling the roller in the contrary direction, when required. The brake is shown at *f, f*, and consists of four wooden brake-blocks, attached by iron shoes to a bar behind them and having rubber packing between the shoes. The screws shown and the handles *h*, are used to operate these brakes. The cranks *m*, working the screws *n*, operate the scrapers *l*, which are used to keep the roller clean in muddy weather. The frame A, is made heavier at *o*, so as to have increased weight there to balance the whole frame-work in turning around. The support *p*, and the guide wheel *k*, might be dispensed with. A great saving in time and in movements hurtful to the road is effected by making the frame circular as described, thus allowing the roller to be turned with the greatest ease. The dimensions are figured on the drawing. A roller of this kind four and one-half feet in diameter, and three and one-half feet long, and weighing some four tons when empty, would cost perhaps $560 to $600; one 5 ft. by 3 ft. 8 in., weighing about five and one-quarter tons (empty,) some $700 to $750. Leaving off the break, would diminish the cost about $50.

Before leaving the subject of macadam top roads, it ought to be mentioned that a bed of rubble stone 10 or 12 in. deep, merely spread uniformly over the road-bed as a foundation, is better than nothing at all, but can never make the same quality of road as the rough paving described above.

The following data are to be used in estimating the cost of the kind of road just described. Rough foundation paving, pieces 5 to 6 in. long, filling up crevices and ramming the whole with hand rammers, costs, after the material has been brought to the

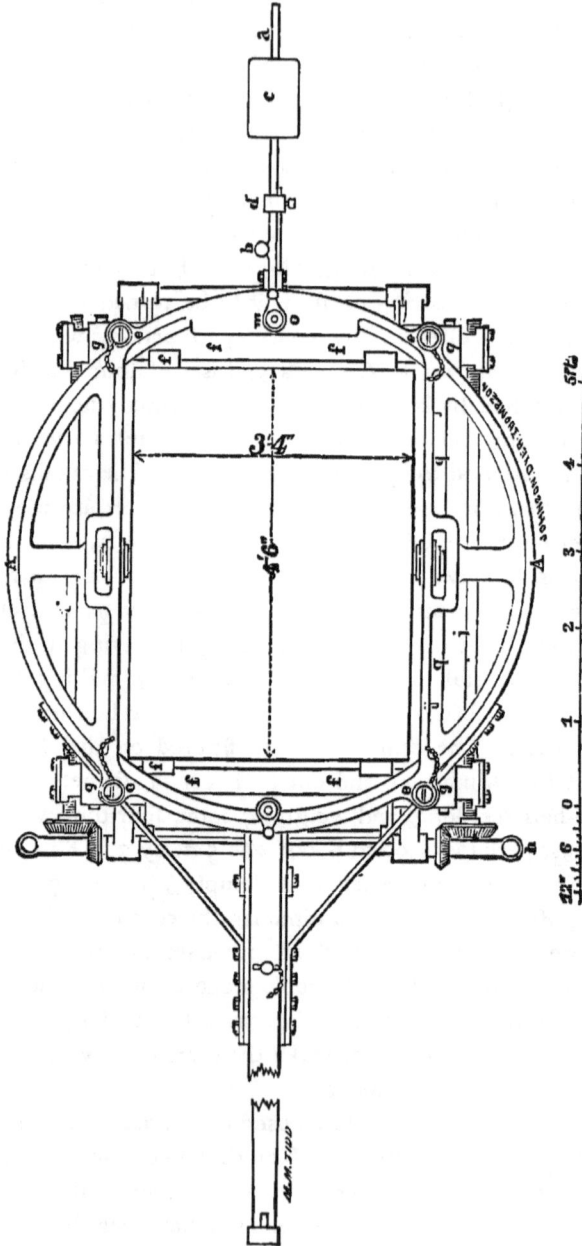

Circular Frame Horse Road Roller Plan.

spot, one day's work of a common laborer for every four square yards, this assuming that the paver gets one and two-thirds common laborer's wages. Same kind of paving if set in sand will cost one day's work of a common laborer for every two and one-quarter square yards.

These figures for the cost of setting rough pavement for a foundation course have been objected to by an American road contractor, as entirely too high, he claiming to set 20, and even 50 square yards to a man per day. An explanation of these different figures probably lies in the phrase, " ramming the whole with hand rammers"; in the general quality of the work done, etc. The writer's own opinion is, that no very fine work is necessary in the construction of the foundation course. Its duties are, to remain pervious, and not to settle unevenly. The same contractor above mentioned wrote, in 1870: " I put down, and keep in perfect order for a year from the time of completion, a 12-inch road (6 to 7 inch foundation, 5 to 6 inch surfacing) at a distance from the quarry of three miles (materials exclusively quarried trap rock) for $1.50 per square yard; wages of men average $2.25 per day, and of transportation, $1.25 per cubic yard. This includes my profit."

To make macadam by hand costs, for sizes from 1¼ to 1½ in. of very hard rock, one day's work for every 0.6 to 0.44 cubic yards, for less hard rock, one day's wages will make 0.7 to 0.6 cubic yards, and of soft rocks 1.76 to 1.17 cubic yards.

In 1872 the estimated cost of crushing stone by the Blake machine ranged from 30 to 60 cents per cubic yard; to crush the same stone by hand, it was estimated would cost from $1.20 to $3.00 per cubic yard.

To spread 14–12 cubic yards of macadam is also about a day's work.

Gravel Top.—Instead of the macadam top described in the preceding articles, screened gravel may be used. These roads are the favorite ones in Central Park, New York, and are probably the best road there is for pleasure drives. It is a matter of some doubt yet whether they do as well for heavy trucking as they do for light vehicles. The foundation for these gravel roads should be the same as the rough paving for the macadam road; some pieces were built in Central Park having a rubble stone foundation, but they are not recommended by their builders. The gravel to be used for the top must be selected with some care; it should be of a hard kind of stone, clean, that is, free from clay, etc., of the right

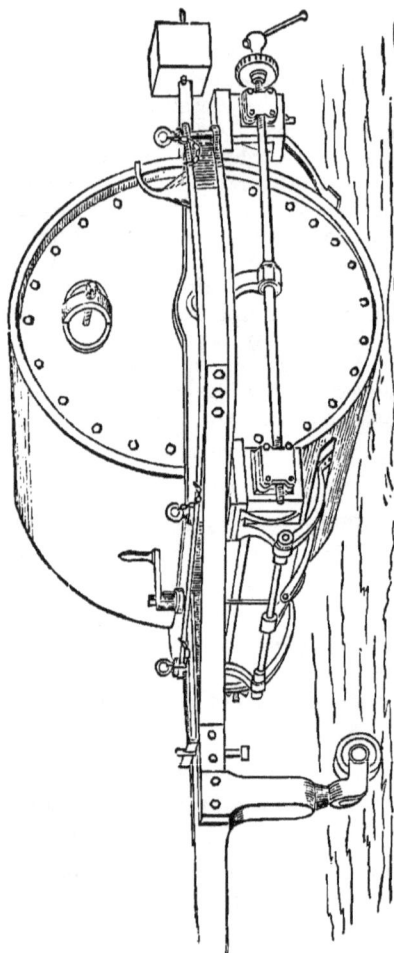

Circular Frame Horse Road Roller Elevation.

color, etc. It is put on in two layers, each rolled, and the top one made compact and firm, by spreading and mixing in some good binding material, sprinkling and rolling. There need be no fear of making a poor road by using the smoothest, most water-worn pebbles, free from all sand, etc., in making a road-top. The upper portions of the river Rhine are remarkable for the clean, smooth pebbles that form its bed to a very great depth. These pebbles are dredged up and used in road-building, making an excellent road-covering at a small expense. There are many miles of such roads in Baden and in the Bavarian Rhine Provinces.

In gravelly soil all the materials that are needed for a good road are frequently on the spot; they only need sorting out and re-laying. For this reason a common gravel sieve often constitutes the principal instrument, whose judicious use will make a good road out of a miserable string of ruts and cobbly elevations. It would be only necessary to sift out and separate the soil under the road to a sufficient depth, into cobbles, coarse gravel, fine gravel and sand; then replace them in the order named and with the proper thickness of layers of each; wet down and roll, and the result would be a good road. As regards the advisability of well constructing roads, the following, from the Bath (Maine) *Times*, of May 11, 1870, is not without instruction. (The Waltham roads therein spoken of are also mentioned in the extracts from a report, which is printed in the Appendix): "I will here submit a comparison of the cost of our road, with those of the town of Waltham, noted for its *good* roads. Waltham has 51 miles of roads; the expense, including everything, of maintaining their highways, except sidewalks, for seven years previous to 1868, was $3,357 per year, or $66 per mile. In 1868, with 60 miles of road, including probably the building of 9 miles, the cost was $6,000, or $100 per mile. The city of Bath has not over 32 miles of roads. The average cost of repairs on our roads for the three past years is $10,153; not including the expense of sidewalk, $317 per mile. At this rate, if we reduce the cost of repairs of our roads to $100 per mile, we could afford to hire money at 7 per cent. and expend $100,000 upon their *permanent* improvement, and it would be vastly cheaper to do so than to continue our present system."

Keeping Roads in Repair.— This subject properly finds its place here, being a matter of skill and a thing of debate only in the case of what we have called foundation roads; pavements and trackway roads, to be considered after this, need no special directions as regards their repair or maintenance.

After a road has been properly rolled, and the surface made

" Barnard Castle " Street-sweeping Machine.

compact and smooth, it should always be maintained in that condition, no matter how great is the amount of travel on it. "A
stitch in time saves nine," here as well as elsewhere. The tendency is to produce ruts; these gather water; this soaks up the
road-bed and spoils the whole. The problem can be put in this
way: To have a good road it is necessary that there be no dust or
mud on the same, and that there be no ruts; therefore, remove the
dust and mud as fast as they are formed, and fill up the ruts as fast
as they are made. The whole matter is here in a nut-shell. It
may be thought, at the first view, that this is too expensive a system. Its principal beauty lies, however, in the fact that it costs
less per mile of road kept one year than the pernicious system of
annual or semi-annual repairs, as will be shown and proved. The
above two rules — sweep off the mud and dust as fast as they are
formed, and fill up the ruts and bad places with new material as
fast as they appear — are all that is necessary to be carried out in
order that there be *continually* a good road. Without continual
repairs, there can be no such thing as a constantly good road —
a proposition that cannot too often be repeated. By repairing a
road annually, or twice a year, it matters not which, the result is,
strictly speaking, a *good* road at no time during the whole year.
The road is wretched just after repairs; it becomes *passable* after a
while, and deteriorates from that day forward, until it is again made
wretched; and so on, *ad finitum*, according to the present, only
too commonly followed system. By the other method is offered
us a road as smooth as a floor, year in, year out, and, let it not be
forgotten, *at a less expense.*

A French engineer, named Tresaguet, was the first, in 1$75, to
call attention to this proper method of making road repairs. His
system — the above described one — was adopted in Baden in the
year 1845, and has been long in universal use in all the active
European countries. The two tables below give, the first, the
actual average quantity of road macadam used per mile of road in
Baden to make the repairs in one year, and show the decrease after
1845. The second gives, in the first column, the cost of materials
and labor required to repair one league for one year according to
the old way, — this column being calculated for the years following

1845 from the cost of the preceding years, and allowing for the increased value of labor and materials, — while in the second column we have the actual cost, as it was with the system followed at the time:

TABLE I.

YEAR.	Cubic yards used per mile in one year to repair roads.
1832, ...	218.6
1839, . . , ...	19S.7
1851, ...	127.2
1855, ...	91.4
1856, . . . , ...	89.4
1860, ...	93.4

TABLE II.

YEAR.	COST OF REPAIRS OF ONE LEAGUE OF ROAD.	
	By old way of so doing, in florins.	By system of continual repairs, in florins.
1835, ...	1,002	1,002
1840, ...	1,086	1,086
1845, ...	1,170	$975\frac{38}{60}$
1850, ...	1,254	$965\frac{40}{60}$
1855, ...	1,339	$835\frac{6}{60}$
1860, ...	1,423	$97S\frac{54}{60}$

These figures are taken as given by the chief engineer of the Baden Public Works, Mr. Keller. He quaintly adds: "These tables give clear evidence in favor of the reduced *cost* by the adopted system. That roads are *better* now than they formerly were, everybody knows." Another German engineer expresses himself to the same effect in a little different way. "It costs no more," says, he, "to keep the roads in repairs now (1864), than it did twenty years ago, when this method (of continual repairs) was not in use, although labor is now three times and materials are twice as dear as they then were." There seems to be no doubt of the superiority of the continual repair system in every respect, producing very much better roads, and at the same time costing less. It need only be tried with us to be thenceforth adopted.

How to Repair Roads on the Continuous System.

We suppose the material for the road covering to lie in regular measured heaps, all ready to be used, at the storage places, once or twice above mentioned, as being 200 to 250 feet apart alongside of the road, but not encroaching upon it. Then for every two or three miles of road, a so-called road-keeper is employed to do the necessary work and repairs. An enumeration of his duties will comprise at the same time an essay on the art of road repairing.

1. The road-keeper is to remove the dust formed in dry weather by sweeping with a brush broom. This is done to greatest advantage just after a slight shower. In muddy weather it is essential that the mud be removed by means of brooms or hoes. A little mud on the surface causes ruts, and much mud softens up the whole road surface. The mud is to be raked up in heaps alongside of the road, there left to dry and then carted off. To hinder as much as possible the formation of any mud, the surface drainage must remain unimpaired; should it be out of order, the water standing on the road is to be swept off. To diminish the wear of the road in dry times, the road should be sprinkled.*

2. Inasmuch as the covering gradually wears off, notwithstanding all precautions, it must be renewed, and should be so renewed gradually, in the same measure as it wears off. The best time to put on new road metalling is during continuous wet weather.

3. In filling up holes, the bottom of the same is to be swept clean of mud, then filled up level with the remainder of the road, not in a heap so high above it as to obstruct travel.

Every care should be taken to have the new material join as speedily as possible with the old portion of the road, and it should be so well laid that it will give the least possible hindrance to vehicles, which will then not avoid the patched places.

4. When many ruts occur in a short distance, the deepest only are to be filled at first. After the patching in these has become solid, then the rest are to be attended to. Long ruts or wheel

---

* Bowles, in his book, "Our New West," mentions the case of the stage road from Sacramento to Virginia City, *via* Placerville, one hundred and fifty miles long, and having an annual traffic of seven or eight thousand heavy teams, and whose proprietors found that the simplest and cheapest way of keeping it in repair during dry weather was to sprinkle the whole of it, — one hundred and fifty miles of mountain road.

tracks are not to be filled up the whole length at once, but only short pieces at a time. If this precaution is neglected, vehicles avoid such places, and new ruts are formed elsewhere.

5. Inasmuch as more material is worn off in a dry season than can be put on, there are then, when wet weather comes, large places to be repaired. These must be mended by degrees, never filling up a piece larger than 8–10 x 4–7 feet at a time, and not having these pieces too near together; when these have become solid, then some more may be fitted in and so on till the whole is done.

Should it however become absolutely necessary to repair a piece of road in dry weather, the place where the new macadam is to be deposited must be loosened up with a pick, then the new material put on and a solid top formed by the judicious use of stone dust or other binding material and sprinkling with water and pounding down with the shovel, or by what may be called " puddling" until the whole be solid. Should a frost or very dry weather occur immediately after macadam has been put on the road in wet weather so that the same will not join on the rest of the road surface, the whole must be removed, cleaned and returned to the storage heaps for future use. A layer of macadam over the whole road should never be put on without treating it immediately afterwards in the manner described above for building new roads, that is, mixing in binding material with the top course and rolling it in wet weather, or after sprinkling.

The road-keeper is naturally also the person to see to the proper delivery on the part of the contractors, if such there be, of the road material in the prescribed places, and to attend to the measuring of the same.

In short and to sum up, it is his business to keep the road in good order, and with proper men and surveillance the desired result is achieved easily and at a less cost, than by any other system. The quantity of macadam required to keep a certain length of road in repair varies very much; it depends, as we have seen, on the care with which the repairs are made, naturally also on the kind of stone used and on the amount of travel over road. For a width of road = twenty feet, the average quantities required per year to keep a length of ten feet in repair, on the system of continuous repairs, has been given as follows:

		Cubic ft.	Cubic yds.
1. Good material and heavy travel		15–20 =	.55–.74
2. Good material and medium amount of travel.		10–15 =	.37–.55
3. Good material and light travel		5–10 =	.18–.37
4. Medium material and heavy travel.		20–25 =	.74–.92
5. Medium material and medium amount of travel.		15–20 =	.55–.74
6. Medium material and light travel		10–15 =	.37–.55
7. Third rate material and heavy travel.		25–30 =	.92–1.01
8. Third rate material and medium amount of travel.		20–25 =	.74–.92
9. Third rate material and light travel		15–20 =	.55–.74

These are the quantities as given by one authority, but from a comparison with the amounts actually used during a period of ten years on thirty-nine roads, having very various amounts of travel upon them and being repaired with all kinds of road metal, it would seem that the foregoing figures are very ample.

The exact relation between the quantity of road material that is necessary to keep a road in repairs, and the amount of travel over it, is still a matter of intelligent observation and discussion. The quantity required does not seem to be proportional solely to the amount of travel, even with one and the same kind of stone used on the same road; as will appear also, when it is considered that were there no travel over the road at all, the surfacing would, nevertheless, wear out by the action of the frost, the rain, etc. As recent an article as the "Annales des P. and C," 1877, p. 226, is devoted to this subject, and does not arrive at any definite general conclusion.

REPAIRS OF MACADAMIZED AND MUCH FREQUENTED STREETS IN CITIES.

In this case, where the amount of travel in one day is often greater than that of a month or more on the town road, the system of continuous repairs ceases to be the best available, on account of the incessant throng of vehicles not giving any repaired place a chance to become solid before it is again ploughed up and scattered. Thus in the city of Paris on the Boulevards, etc., the continuous system has been abandoned and the practice now is to let the street gradually wear down three to four inches, then close half of it (divided "fore and aft") to travel, loosen it all up with picks and put on a layer three or four inches (best not to put on more than that), spread a thin layer of sand over this, sprinkle and roll heavily.

It often happens that the men put too much of the sand on; in that case, the road, after it is all done, is finally well watered and the roller again passed over it a number of times. This operation causes the superfluous binding material to come to the surface in the shape of thin mud and leaves the road covering as hard and smooth as mosaic, making a most excellent driveway. It emits a sonorous, ringing sound on being driven over and remains clean and without mud throughout the heaviest rain-storms. The rolling of the streets in Paris, is done by a company owning a large number of steam rollers; in paying them for work done, the city was obliged to go back to first principles for a measure of such work, it being found impossible to estimate correctly by the square measure of surface rolled to such and such a degree of hardness. The measure adopted is that of weight multiplied into the distance it has been moved, or " feet pounds " as we should say. It has been found from many years experience that to roll one cubic meter of macadam requires 4–5 " Kilometer-tonnes," and this is true whether the layer of macadam be three and one-quarter or ten inches thick. Expressed in our measures this is 11,020–13,775 feet tons @ 2,000 lbs.=2.09–2.61 mile tons per cubic yard of macadam.

The advocates of the steam road-roller claim, that by means of that machine, they are enabled to make a road that will wear out evenly and uniformly for 4 or 5 inches, so that the operation of patching need never be resorted to. The steam road-roller can also be used for " picking " up a road, for which purpose the roller is armed with sharp spikes, and is then driven over the surface to be " picked " up.

PAVEMENTS AND TRACKWAYS.

No essay on roads would be complete without some mention of these two species of road surface, though the use of the former is confined principally to streets, and that of the latter is out of date.

Pavements are either of stone, wood, iron, various concretes, asphalt, and may be of still other substances.

Stone Pavements.—The modern sizes of paving stones may be seen from the following cases. The Boston size is 4½''×3½'' ×7'' deep; New York Belgian, 6–8''×5–6''×6–7'' deep; new Broadway pavement, also called Guidet pavement, 3½–4½''×10– 14''×7½–8½'' deep. This last is laid with the long sides of the stones across the street; and, as far as the author's judgment goes, is the best size for stone pavement there is. The Boston size is too small, and allows of no bond between the separate paving stones. Further, the weakest part of each stone being its edge, it follows that the more edges there are in a given surface of pavement, the speedier will it wear out, each stone becoming rounded and slippery. It is only the excellent workmanship and great care displayed in setting these stones in Boston that prevents these facts from being at once apparent to all. When it is added that in setting pavements, the natural soil, except it be sand or fine gravel, is in all cases to be excavated 12–19 inches, and then filled up 5–12 inches, according to the solidity of the subsoil, with *clean*, coarse sand or fine, *clean* gravel, and the paving stone set in this and well rammed down with hand rammers, about as much is said on this topic as can be said without going into long details.

From four and one-half to six cubic feet of sand are required for every square yard of paving. In setting two different pavements, the same written rules may be exactly followed in either case, yet one be much better than the other, so much depends here upon good, careful, conscientious workmanship.

Wooden Pavements.—There are so many kinds of these, that it would be out of place to enumerate and describe them here. Their advantages are, less wear on tires and horses, less noise and smooth traction; a disadvantage, is their slipperiness in the winter. There seems to be a sort of notion that wood pavements and coal tar must go hand in hand; but there certainly is no necessity for this. Coal tar is applied as a preservative to the wood; but it must be acknowledged that many better ones are known and indeed are used, to the utter exclusion of coal tar, in all cases where it is desired to preserve wood, except in this of wood pavements. No wood should be used in paving that has not been first subjected to some approved method of preservation, or impregnation, as it is frequently called. The best manner of setting the same is still a mooted point, which it would be presumptuous at present to decide.

A valuable contribution to the subject of wooden pavements, is the report of the Commission appointed by the city of Boston to consider this subject, in 1872, City document, No. 100, 1873. The Commission come to the conclusion, that the best way to preserve the wood that is put down, is by the method called Burnettizing, after its inventor, Sir H. Burnett, of England, in 1838. It consists of treating the wood to be preserved with chloride of zinc. The Commissioners wisely add: "Your Commisioners are of the opinion that if the city adopts *any* method of preserving blocks to be used for pavements, some additional security should be had that the treatment of the wood shall be thorough and complete." As regards the construction of the pavement, the Commissioners recommend spruce blocks (for this section of the country), lay stress on the necessity of a solid, uniformly constituted, and rolled gravel foundation, and then say: "The rows or blocks should be set square across the street, and should be about 4 inches thick at top, with spaces of about one-half inch between the rows. This may be done with blocks of uniform thickness set apart, or with tapering blocks half an inch thicker at bottom than at top. The latter arrangement is the more costly, but it is believed by some that it will stand better, by reason of its covering the whole surface of the foundation. Longer trial is necessary to settle this point beyond dispute. Blocks of only a short chamfer at the top leave the interspace too narrow, as the blocks wear down." The Commission named consisted of "two chemists, two practical mechanics, and one civil engineer."

Cast-iron pavements are out of favor on account of their great cost, and concrete pavements are a matter of experiment as yet.

Asphalt pavements are chiefly used in Paris. They are slippery in wet weather, and produce a very disagreeable, penetrating dust in dry weather. It is necessary to prepare a bed of macadam to lay them on, and they are not used in Paris except in streets where the gas pipes are carried either in the sewers or under the sidewalks, as any leak of gas would destroy them. Their use is a matter of doubtful economy.

Trackways are, as has been mentioned, out of date. Where a common road does not suffice now-a-days, a railroad is built; but time was when trackways were of considerable importance. They consist, if of stone, of large flat stones, say 12'' deep and 4–6 feet long by 14''–16'' wide, solidly bedded in two parallel rows, at such distance apart as to make of each row a track for the wheels. The space between is paved. They are of course very expensive, but cost little to repair, and enable a horse to pull a very great load. As has been mentioned, Telford made use of such a trackway, to avoid cutting down a hill, on his Holyhead road. There were two hills, each a mile in length, with an inclination of 5 in a hundred. It would have cost $100,000 to reduce this grade to 4⅙ in a hundred, but nearly the same advantage, in diminishing the tractive force required, was obtained by keeping the 5 in a hundred grade, with moderate cuttings and embankments, and making stone trackways, at a total expense of less than half the former amount.

"Plank roads," once so much in vogue in the United States, may not improperly be classed among roads with trackways, and, with them, also among the things that were. From their perishable nature, they can never advantageously do more than help the development of a new country, and in this, as well as other States, are yearly becoming more and more impracticable on account of the constantly increasing price of lumber.

On the Resistance to Motion or the Force Required to Move Vehicles on Different Kinds of Roads.

Before, as well as since the introduction of railways, engineers in England, Germany and France made many experiments on the force necessary to pull different vehicles, at various speeds over various surfaces. To enumerate the details of all these experiments would be perhaps useless; a few general results only are here given.

Experiments, as above indicated, were made by Edgeworth, Count Rumford, Bevan, Macneill, Minard, Navier, Perdonnet, Poncelet, Flachat, Morin, Kossak, Umpfenbach, Gerstner, and no doubt others, a list of authorities that proves the subject to have been well nigh exhausted. The experiments of Morin, made in 1838–41, appear to have been made with a degree of care and accuracy, leaving nothing more to be desired, and the following table is an extract from his results,* and gives that fraction of the weight of the vehicle and load, which is required to move them on a level road:—

---

* A full account of Morin's experiments on the resistance to motion of vehicles, on the wear caused by different vehicles on roads and on the loads different vehicles shou'd carry so as to produce the same wear, may be found in Morin, *Expérience sur le tirage des Voitures, Paris,* 1842.

Character of the Road.	2-wheeled carts.	Trucks, 4 wh., three and four horse.	4-horse stage-coaches, on springs.	2-horse carriages, body on springs.
Firm soil, covered with gravel 4''–6'' deep,	$\frac{1}{12}$	$\frac{1}{9}$	$\frac{1}{8}$	$\frac{1}{8}$
Firm embankment covered with gravel 1¼''–1½'' deep,	$\frac{1}{16}$	$\frac{1}{11}$	$\frac{1}{10}$	$\frac{1}{10}$
Earth embankment in very good condition,	$\frac{1}{41}$	$\frac{1}{29}$	$\frac{1}{28}$	$\frac{1}{28}$
Bridge flooring of thick oak plank,	$\frac{1}{70}$	$\frac{1}{46}$	$\frac{1}{41}$	$\frac{1}{42}$

BROKEN STONE ROAD.			walk	trot	walk	trot
In very good condition, very dry, compact and even,	$\frac{1}{75}$	$\frac{1}{64}$	$\frac{1}{48}$	$\frac{1}{41}$	$\frac{1}{49}$	$\frac{1}{42}$
A little moist or a little dusty,	$\frac{1}{53}$	$\frac{1}{38}$	$\frac{1}{34}$	$\frac{1}{27}$	$\frac{1}{34}$	$\frac{1}{27}$
Firm, but with ruts and mud,	$\frac{1}{33}$	$\frac{1}{24}$	$\frac{1}{21}$	$\frac{1}{18}$	$\frac{1}{22}$	$\frac{1}{19}$
Very bad, ruts 4''–4½'' deep, thick mud,	$\frac{1}{19}$	$\frac{1}{14}$	$\frac{1}{13}$	$\frac{1}{10}$	$\frac{1}{12}$	$\frac{1}{10}$

Good pavement,						
Dry,	$\frac{1}{90}$	$\frac{1}{65}$	$\frac{1}{57}$	$\frac{1}{38}$	$\frac{1}{59}$	$\frac{1}{39}$
Covered with mud,	$\frac{1}{69}$	$\frac{1}{50}$	$\frac{1}{44}$	$\frac{1}{33}$	$\frac{1}{45}$	$\frac{1}{34}$

To take an example, suppose we have a truck weighing with its load 9,000 lbs. How many pounds traction will be required to move the same?

Ans.—On firm soil, gravel 4''–6'' deep, that is, a newly repaired road, as we often find it, ($\frac{1}{9}$ by table), 1000 lbs.; on best kind of embankment, ($\frac{1}{29}$ by table,) 310.3 lbs.; on broken stone road in good condition, ($\frac{1}{64}$ by table,) 166.6 lbs.; on broken stone road, deep ruts and mud, ($\frac{1}{14}$ by table,) 643. lbs.; on a good pavement, ($\frac{1}{65}$ by table,) 138.5 lbs. Or, since the tractive force of a medium horse when working all day is said to be about 125 lbs., we need in the first case, 8 horses; in the second case, 2½ horses; in the third case, about 1¼ horses; in the fourth case, about 5 horses; and in the fifth case, only one good horse to move the same entire load all day.

These facts expressed in the preceding page or two in striking, yet perhaps dry figures, can be nearly as well given in popular language.

Says a correspondent (Dr. Holland), of the *Springfield Repub-*

lican, writing from England, after describing the kind of horses in use there:—

"Now with all these horses the rule follows that every pound of muscle does just as much work on the road as two pounds do in America. The cab and omnibus horse does twice as much as the same horse does in America. The draft horse does as much at the dray as two ordinary dray horses in America, and the little horses, which are driven mainly in butchers' carts and grocers' carts, will tire a cab horse to follow them with no load at all.

"In connection with these statements it should be recorded that the speed of all vehicles in the streets of London, whether the localities be crowded or not, is at least a third faster than it is in corresponding streets in American cities. The ordinary speed of vehicles in London, in which passengers or light loads are transported, is one which is considered not entirely safe in Main street, Springfield, Mass., and one which, in some streets of Boston or New York, would be at once checked by the police. A man who sits in a 'hansom' finds himself driven at an unprecedented pace through crowded thoroughfares, and Yankee though he may be, he will often wonder whether he is going to bring up at last without a broken neck.

"I mention this matter of speed, particularly, because it shows that even more work is done by one horse in London, than by two in New York. He not only draws as large a load, but he travels with greater rapidity. The streets of London present such a spectacle of headlong activity as no American city can show, in consequence of the rapid passage of all sorts of vehicles through the streets. I might add to this statement, touching the superior speed of the London horses, a word about the greater weight of the carriages which they are obliged to draw behind them. All carriages are built more heavily in Great Britain than in America. They are built to last, and many of them seem to me to be superfluously heavy.

"The point which I wish to impress upon my American reader is simply this:—that the English horse, employed in the streets of a city, or on the roads of the country, does twice as much work as the American horse similarly employed in America. This is the patent, undeniable fact. No man can fail to see it who has his eyes about him. How does he do it? Why does he do it? These are most important questions to an American. Is the English horse better than the American? Not at all. Is he overworked? I have seen no evidence that he is. I have seen but one lame horse in London. The simple explanation is that the Englishman has invested in perfect and permanent roads what the American expends in perishable horses that require to be fed. We are using to-day, in the little town of Springfield, just twice as many horses as would be necessary to do its business if the roads all over the town were as good as Main street is from Ferry to Central. We are supporting hundreds of horses to drag loads through holes that ought to be filled, over sand that should be hardened, through mud that ought not to be permitted to exist. We have the misery of bad roads, and are actually or practically called upon to pay a premium for them. It would be demonstrably cheaper to have good roads than poor ones. It is so here. A road well built is easily kept in repair. A mile of good macadamized road is more easily supported than a poor horse."

Other results of Morin's experiments are as follows:

1. The force required to draw a vehicle, is directly proportional to the load, and inversely so to the diameter of the wheels; in other more common words, the tractive force increases in the same ratio that the load increases, and the diameters of the wheels decrease.

2. On a paved or well built macadam road, the tractive force is independent of the width of the tires, provided the same is more than three or four inches. On compressible roads, such as new gravel, on a meadow, etc., the tractive force diminishes with an increase in the width of the tires.

3. Other circumstances being equal, the tractive force is the same for vehicles with and without springs as long as the horses are not moving faster than a walk.

4. On paved and well macadamized roads the tractive force increases with the velocity, according to the law, that beyond a velocity of 2¼ miles per hour (3.3 feet per second) the increase of the tractive force is in direct proportion to the increase in velocity;

this increment is however less, the softer the track or road and according as the vehicle is best provided with springs.

5. On soft earth embankments, or on sand or sods, or on streets newly covered with gravel, the tractive force is independent of the velocity.

6. On a well-made pavement of regular shaped stone, the trac-tive force, horses on a walk, is about three-fourths of that on a good macadam road, but with horses on a trot, the two are about equal.

7. The wear on the road is greater the smaller the diameter of the wheels and greater in the case of vehicles without, than for those with springs. Most road-rollers, as now in use, have twc small a diameter besides being two light and consequently do not properly compress the road surface.

8 The tractive force, as well as the wear on the road, is greater in the case of vehicles that have their wheels placed at an angle with the vertical by reason of the ends of the axle-trees being bent down, than for those that have their wheels set plumb and the cen-tre line of the axle-trees level.

PART II.

ON THE "BEST METHODS OF SUPERINTENDING THE CONSTRUCTION AND REPAIR OF PUBLIC ROADS IN THIS COMMONWEALTH" (MASSACHUSETTS).

In looking for a solution of this question the people of the Commonwealth man turn as they choose, either to the West or to the East, to see a guiding star; to the city of Chicago, or to the city of London, both under a republican form of government, alike or similar to that we live under. It lies in the establishment of a Board of Works, composed of a number of able men, well paid for their services, gradually changing in their membership in the Board who shall have this and only this as their occupation, and who can therefore be held responsible for their acts. This is the system that has been adopted both in London and in Chicago and with remarkable success and resultant benefits. There are many other systems in use in foreign countries all of which however seem to be inapplicable here, placed as we are, under so different forms of government; hence, though well acquainted with the systems adopted in France and in Germany, the writer has not described them here.

The history of "the Metropolitan Board of Public Works of the City of London" is about as follows:

What is known as the city of London consists in reality of a great number of what we should call towns, there called parishes, and of which the "*City of London*" is only one single member. Each one of these parishes had, and still has in most respects, its own local government, and in consequence took care of its drainage, its streets, etc., etc., as seemed best and as it liked, some better, some worse, and some not at all. This state of things in the matter of drains and sewers finally led to a most deplorable condition of affairs; there was not nor could there under these conditions be such a thing as a *system* of sewers, and consequently a proper and adequate drainage; the death-rate increased to an alarming extent and matters came to be universally regarded as past all endur-

ance. What could be the remedy? No well grounded complaint
could be made against the majority of the men composing the various local governments, since they were good and honest citizens,
and hence no change in the separate goverements could ever bring
relief. The fault lay not in the men, but in the system of ruling
they were called upon to fulfill, that is, in the incompetent and
faulty treadmill of government they were annually called upon to
keep in its usual operation. It was then seen that by having an
elected power to supervise and regulate the sewage affairs of the
whole metropolis, a complete *system* of drainage could be carried
out, and thus only. Such a regulating power is exercised by the
metropolitan Board of Public Works, chartered by Act of Parliament and composed of members elected from all parts of London.
It is perhaps in place here to explain what is meant by a *system* of
sewers as the same definition will hold good in other matters; as
for a *system* of roads, of drainage and irrigation of lands, etc.
Perhaps the best illustration would be to refer one to the veins and
arteries in the human body, or to the body of a tree, from its trunk
through the branches growing smaller and smaller down to the
smallest twig that may be on it. It will be at once seen how different any arrangement, in which may be the wisdom to contrive,
the strength to uphold and the beauty to adorn, like this, is from a
miserable patchwork such as cannot but arise where the separate
parts of one whole are each left to guide themselves without any
unity of action or design, as to their final resultant. The London
Board of Public Works had some extraordinary powers conferred
upon it, such as the right to levy assessments on real estate bene-
fitted by their improvements, and others. Originally constituted
merely to plan and execute a system of sewerage for the metropolis, this Board of Public Works soon showed itself so useful and
beneficial in its actions that other matters were placed in its charge,
such as the laying out of new streets, the building of the Thames
embankment,—a work of exceeding great magnitude and importance,—and there seems to be no doubt that in all public works
London will find it advantageous to employ its Metropolitan Board
of Public Works.

In the city of Chicago there has been a Board of Public Works

almost from the very start. It arose there from the union of the
water supply and the sewerage commissioners, and has existed
since May, 1861. No less than in London, it has proved to be of
great benefit to the community; and it would have been impossible,
under any other system, to have executed in so satisfactory a manner the many and useful public works for which Chicago is famed.
At the risk of introducing in this place some very dry reading, a
general synopsis of those parts of the city charter which relate to
the Chicago Board of Public Works is here given. The whole
may be found in a copy of " Laws and Ordinances, Chicago, 1866:"

SEC. 1. Establishes a body known as the " Chicago Board of
Public Works," to consist of (3) three members, chosen by the
people, one from each division of the city.

The first *three* chosen for one, two and three years; after that,
one each year for three years.

SEC. 2. Each member of board shall receive annual salary of
three thousand dollars (by Act of February, 1866); give bonds for
faithful discharge of duties; pay over all moneys, papers, etc., at
expiration of his term, or when ordered by city council.

SEC. 3. Board to elect president and treasurer, and make by-
laws.

SEC. 4. Majority constitutes quorum; records to be kept of proceedings; copies of all plans, estimates, etc., to be kept; report
(annual) to be rendered on or before   each year,
or when required by city council. Each member authorized to
administer legal oaths.

SEC. 5. Board shall have special *charge* and *superintendence*,
subject to the laws and ordinances of the city council, of all streets,
lanes, alleys, etc., in the city of Chicago, and of all walks and
crossings in the same, and of all bridges, docks, wharves, public
places, landings, grounds and parks in said city, and of all halls,
engine-houses, and other public buildings in the city belonging
to city, except school-houses, and of the *erection* of *all* public
buildings; of lamps and lights in streets, etc., and in public buildings, and repairs of same; of the harbor works and improvements;
of the city sewers and drains and of the water works; of the fire-
alarm telegraph, and all public works and improvements hereafter

to be commenced by the city, as well as such other duties as may be prescribed by the city council by ordinance.

SEC. 6. All applications or propositions for improvements or new works of kind specified in section five, shall hereafter be first made to Board of Public Works, or if made first to city council, shall be by them referred to Board. Upon receiving application, Board shall investigate the same, and if they find such work necessary and proper, shall thus report to city council, with an estimate of the expense thereof. If they do not approve of such application, they shall report the reasons for their disapproval, and the city council may then in either case, reject said application or order the doing of work or making of public improvement, after having first obtained plans and estimates thereof. The Board may also in like manner recommend whenever they think proper, any improvement of the nature above specified, though no application has been made therefor.

SEC. 7. Shall be duty of Board to procure for city full plans and estimates of contemplated improvements, when so ordered by council.

SEC. 8. Whenever any public improvement shall be ordered by city council, and money appropriated, Board shall advertise for proposals for doing work; plans and specifications of same first placed on file in office of Board, which plans and specifications shall be open to public inspection; advertisement to state work to be done, and to be published ten days at least. The bids shall *be sealed bids*, directed to board, and accompanied by bond to city, signed by bidder and two responsible sureties, in sum of two hundred dollars, conditioned he shall do work if awarded to him; in case of his default to do so, etc. Bids to be opened at time and place mentioned in advertisement.

SEC. 9. All contracts shall be awarded to lowest reliable bidder, and who sufficiently guarantees to do work under superintendence and to satisfaction of Board: *provided*, that the contract price does not exceed the estimate, or such other sum as shall be satisfactory to Board. Copies of contracts to be filed with city comptroller.

SEC. 10. Board reserves right, in contracts, to decide questions as to proper performance of work and meaning of contracts; in case of improper construction may suspend work and relet same, or order entire reconstruction; or may relet to other contractors and settle for work done, etc.

In cases where contractor properly does work, Board may, in their discretion, as work progresses, grant to said contractor estimate of amount already earned, reserving fifteen per cent. therefrom, which shall entitle holder to receive amount, all other conditions being satisfied.

SEC. 11. In case prosecution of any public work be suspended, or bid be deemed excessive, or bidders be not responsible, Board may, with written approval of treasurer, where urgency of case and interests of city require it, employ workmen to perform or complete any improvement ordered by council: *provided*, that the cost and expense shall in no case exceed the amount appropriated for the same.

SEC. 12. All supplies of materials etc., when costing over five hundred dollars, to be purchased by contract, subject to same conditions as letting out work.

SEC. 13. Whenever Board think necessary for interests of city, to protect same from damage or loss, shall report thus to aldermen, and reasons for same, asking power to give contracts without notice required above, and aldermen may grant request: *provided*, three-fourths vote for it.

SEC. 14. Whenever Board is of opinion work may be better done without contract, shall so report to council, and same may authorize Board to procure machinery, materials, etc., hire workmen, etc.: *provided*, a three-fourths vote be in favor of granting authority.

SEC. 15. All contracts and bonds by Board to be in name of city.

SEC. 16. No member to be interested in any contract; all contracts made with any member interested, city may declare void; any member so interested shall forfeit his office and be removed therefrom; the duty of every member of Board and of every officer of city to report delinquency, if discovered.

SEC. 17. All existing contracts executed by city, by water or sewerage department, etc., to be carried out by Board.

SEC. 18. Board shall nominate each year the various officers, now provided for by ordinance, which serve in the departments under their special charge, the city engineer, superintendent sewers, streets, etc. Shall be empowered to employ from time to time such other superintendents, clerks, etc., as they may deem necessary, subject to ordinance as regards pay, etc.

SEC. 19. Board to have charge and superintendence of works made for city, and paid for by private individuals or by State. Plans for same to be approved by Board.

SEC. 20. Board shall, on or before   every year, submit to auditor, by him to be presented to council with annual estimate, statement of the repairs and improvements necessary to be undertaken for current year, and of the sums required by Board therefor; report to be in detail; report, having been revised by council, sums required shall be provided for in annual tax levy. All moneys to be paid to any person out of moneys so raised, shall be certified by president of Board to auditor, who shall draw warrant on treasurer therefor, stating to whom payable and to what fund chargeable; such warrant to be countersigned by president of Board.

SEC. 21. Board to keep accounts showing moneys received and spent, clearly and distinctly, and for what purpose. Accounts to be always open for inspection of auditor or any committee appointed by city council.

The object of introducing this synopsis here has been to give a complete picture of just what such a Board of Public Works is. It will be seen upon a little examination how entirely different a thing it is from the usual and only too customary "committee." Perhaps the greatest fault of a committee is its entire lack of what might be called body *and* soul. If corporations, as has been said, have no souls, a committee may be said to have neither body nor soul. It is alive to-day, wields great power, decides vital and important questions, and yet is nowhere to-morrow, and seemingly even its component atoms have vanished from the face of the earth. It is amusing and yet sad, when the action of some such committee has caused trouble to read some time after, that it all "is exceedingly discreditable to whoever is responsible for it."

How much better to have a conservative, expert and reliable body, the members of which have *no other business* than to attend to their duties as such, who are well paid for it and consequently can at any time be held strictly responsible for their actions. With such a power, wisely governing and regulating the roads of this Commonwealth, it would be an easy matter to make thorough improvements in the legislation concerning roads and in the roads themselves.

These are two changes the need of which is generally felt at present and has found expression in various ways.

It may be well to quote one at least, notable for saying very much in little compass,—of these calls for improvement, in this connection, and adding some more as belonging to this subject in the form of an interesting appendix. Says Gov. Claflin in his Inaugural: "Few things are of greater importance to a community, or a surer test of **civilization**, than good roads. Those of our citizens who have visited Europe are unanimous in the opinion that our public roads are far inferior to those of other countries, where the means of easy and safe communication are better appreciated. The science of road-making is apparently not well understood; or, if it is, the present modes of superintending the construction and repair of roads are so defective that the public suffers to an extent of which few of us are aware. It may be found upon investigating the cause of our miserably poor and ill-constructed roads, that the laws relating to this subject need revision, so as to give more uniformity in their construction and the repair of our highways. It is evident, also, that the science of road-making should have a prominent place in the course of applied mathematics at the Massachusetts Agricultural College."

We stand then in this matter of roads at precisely the same point that the good people of London did ten or a dozen years ago in the matter of their drainage, and our remedy is the same. The fault lies in the machinery of government; originally built up to cater to the wants and needs of a newly settled country,— a colony breaking a path through the wilderness,—it has long since ceased to satisfy the demands of the present *State* in no matter so essentially as in that of its government and laws relating to common roads and highways. This is a subject requiring special knowledge, to be acquired only by long experience or the shorter method of imbibing the experience of others, which, on analyzing it, is all that any *study* amounts to; formerly it was not so, and most any one sufficed to make improvements on Indian paths. We need then an *expert* government on this point.

There should be a distinction made between first, second and third class, or between, as they might be called, State, County and Town roads; the first two should not be left to be dealt with as it is the pleasure of each town. A chain cannot be perfect unless every link in it is so; no more can a road. The State must attend to the State and County roads and set a proper example at least to be followed by the towns in the case of their roads. We need then a higher power than that of the towns.

It has been previously shown how we need a power that can be held responsible and is somewhat permanent, and to put it all together, we need, to order and maintain our highways, a Massachusetts Board of Public Works. For some years it would have its hands full in improving the existing main roads and laying out some new ones, but in course of time, as in the older countries of Europe, its principal business would be the *maintenance* of the roads. It must be remembered that the Board of Public Works is merely the intelligent servant and adviser of the legislative and executive; whatever sums the legislature appropriates for certain objects, that is taken by the Board and made to yield its most in the shape of work accomplished. Beyond this and keeping its accounts, it has nothing to do with money or taxation.

The small state of Baden, a part of Germany, has been heretofore mentioned as a model in road construction and the care of the same. From a brief history of the roads of that country and their present management, we may take some useful notes. The account is that of the Chief Engineer of the department of "Roads and Hydraulic Engineering," which has this matter in charge and is therefore reliable.

"In Baden the condition of the roads has been a subject of great care. Within the last forty-five years many millions have been spent upon them, and experience has shown this expenditure

to be one of those most advantageously spent. As most of the
roads are well laid out, and as there are plenty of them, there
remains now (1863) mainly the keeping in repair of the roads to
be attended to and not to build any new ones. Our endeavor now
is, to do this at the minimum of cost. Statistics gathered on this
subject, show good results and point out to us the means of arriving at still better ones. The present road law was made in 1810.
That part of the old law which relates to the maintenance of roads
is still in force, but that part requiring labor as a road-tax was
abolished in 1831, and likewise most of the road police regulations.
The appropriation for roads had to be increased 250,000 florins to
pay for the abolished road-tax labor and to make up 170,000 florins
previously received from tolls, which were also abolished in 1831.
The system now is as follows: All town roads are taken care of by
the towns. The State merely appoints and pays a road-master, so
called, who superintends fifteen or twenty road-keepers and reports
on the state of the roads, the reasons for their bad condition, if that
be the case, what is needed, etc. The law for second class or
county roads was formerly, that when they were of importance to
several towns, they had all to help maintain the same. As this
gave rise to continual bickering and quarreling, in which the road
suffered most, it was changed in 1856. They are now taken care
of under the direction of the State and paid for partly by the State
and partly by the towns in which they are situated. Most of the
roads under this head are those which have risen in importance
since the building of railroads, and are generally those that lie perpendicular to the direction of the railroad they are influenced by.
The towns not having the means very often to properly improve
and repair such, it was found necessary and expedient to give them
the aid of the State, and in order to procure the necessary funds,
all roads that run parallel to railroads and all those that had lost
their importance by the construction of railroads, were in 1855
stricken from the list of state roads. These latter as the name im-
plies, are wholly under the care and kept up at the expense of the
State.

In 1835, the total length of the State roads was .   .   1,430.8 English miles
In 1855,   "   "   "   "   .   .   1,500.8   "

In 1855, by excluding several State roads, this last length   English Miles.
    was reduced to . . . . 1,142.4   "
In 1861, it had increased to . . . .   1,190.0   "

Second class Roads (keeping partly paid for by State.)
In 1835, the length of these was . . . . . 467.6 English miles.
In 1861,   "   "   . . . . . . 630.0   "

So that the State had, in 1861, in all, 1,820 English miles of road to maintain, the towns helping to pay on six hundred and thirty miles thereof.

The areas, population, and population per square mile of Baden, Prussia, France, Hanover and Massachusetts, according to recent census, are as follows:

Country.	Year.	Area, sq. miles.	Population.	Pop. per sq. mile.
Baden.	1871.	5,891.	1,461 562.	243.
Prussia.	1871.	134,045.	24,643,698.	184.
France.	1872.	204,088.	36,102,921.	177.
Hanover.	1871.	14,857.	1,963,080.	132.
Massachusetts.	1875.	7,800.	1,651,912.	212.

Baden did have, at a time when her population per square mile was less than it is now, and Prussia, France, Hanover, and many other countries that could be named, have now got, and for the past 40 or 50 years have had, a system of common road management and resultant common roads, of the character above described; while Massachusetts with a population of 212. per square mile, and corresponding wealth, and others of the States of the Union, have a species of highway management, and *its* resultant and corresponding sort of highways, which, in thinking of the roads of the countries named, are but as evidences of a partial civilization.

" The statistics of the road repairs are kept in the following manner. The road-keepers are required to keep a record of all draught animals that pass in either direction. Horses that are being ridden, animals not before a vehicle, and teams going to and from the fields, are not counted. These records are kept only during the working hours. Likewise, not during the whole year, but only four months in each year, so selected as to give an average amount of travel. The travel on the road on Sundays and out of working hours is taken from a few observations; it is a very small percentage of the whole. At the end of the year these records and observations are collected and graphically represented on a map of the whole State. The different roads are drawn of a different thickness of line, according as the amount of travel on them is greater or less. The quantity of road metal used per yard of road, and the kind of metal used, give the data for another such map, in which the different colors of the roads represent the different materials used in their repair, and the figures on them and their thickness show the number of cubic yards per mile required to keep the road in order. Finally, we have a third map, which indicates, by the thickness of the several lines representing the roads and by the figures on them, the total cost per mile of repairing the road one year."

With this picture of a country happy and prosperous, in the possession of good and well-kept roads, it may be well to leave the subject.

Massachusetts wants for her proper development, much better roads than she now has; and, reckoning for a period of say fifty years, she can have these good roads, and have them kept in order, at a less cost than that of keeping up the present poor ones for the same time. Besides this, we should see in the one case a healthy state of internal communications and trade; in the other an absence of both. Let each citizen so act and do his part, that these benefits may accrue to the Commonwealth.

APPENDIX.

For the sake of arriving at some practical end, I have requested the gentlemen to whom the prizes for essays were awarded to suggest what form of legislation would be desirable as a change from our present inefficient system of road management, to one which should promise better, more economical and more satisfactory results. The large and varied experience and observation of these gentlemen, all of whom are competent engineers, entitle their opinions and judgment to favorable consideration ; and the following, submitted by them, may serve as a basis or outline for future legislation. C. L. F.

AN ACT FOR THE MORE PERFECT CONSTRUCTION AND MAINTENANCE OF THE COMMON ROADS OR HIGHWAYS THROUGHOUT THIS COMMONWEALTH (MASSACHUSETTS).

SEC. 1. Establishes a body to be known as the State Board of Highways and Bridges, to consist of three skillful civil engineers, or persons practically expert in the science of road-making, to be appointed by the Governor with the advice and consent of the Council, and to have their office in the State House.

SEC. 2. It shall be the duty of the Attorney-General, personally or by his deputy, to give his council and opinion on such matters as he may be called upon by the Board, for which service his compensation shall be

SEC. 3. The first appointment of members of the Board of Highways and Bridges shall be made on or before , and there shall be appointed one member each for the terms of one, two, and three years; after that there shall on or before each year be appointed one member for the term of three years.

SEC. 4. Each member of the Board shall receive an annual salary of dollars; give bonds for the faithful discharge of his duties; pay over all moneys, papers, etc., at the expiration of his term or when ordered by the Governor and Council.

SEC. 5. Board are to elect a president and treasurer, and make their own by-laws.

SEC. 6. A majority of the Board constitutes a quorum; records to be kept of all the proceedings; copies of all plans, estimates, etc.,

to be kept; report to be rendered on or before   each
year, or when required by the Governor and Council.   Each member authorized to administer legal oaths.

SEC. 7. Said Board-shall prepare and submit to the
legislature a plan for the systematic classification of all the highways
and townways in this Commonwealth into two or more of the following three classes:—

Class 1. State roads, to be controlled and maintained wholly by
the State.

Class 2. District roads, to be controlled and maintained by the
State, but the expense thereof to be borne by the towns and cities
of the districts in which said road shall lie, and the State, in such
proportions as said Board shall apportion.

Class 3. Town roads to be controlled and maintained as now
provided by law.

The construction of new roads, of the three classes above specified, to be done as follows:—

Class 1. State roads, to be laid out and built by the State,
through the Board of Highways and Bridges.

Class 2. District roads, to be laid out, etc., by the county commissioners, as now provided, but the board to have the final approval or disapproval of the proposed plans and profiles for said
road, and also to have the charge and superintendence of their construction.

Parties aggrieved by the refusal or neglect of county commissioners to lay out a road, to have the right to appeal to the Board
of Highways.

Class 3. Town roads, to be laid out and constructed as now
provided by law.

SEC. 8. The paying of road taxes by labor is hereby abolished,
and all road taxes are hereafter to be paid in cash.

SEC. 9. Board shall have the special charge and superintendence, subject to the laws and resolves of this Commonwealth, of
all the highways and bridges, and the public works appertaining
thereto, which are or shall be executed or maintained, wholly or in
part by this Commonwealth.   They shall also perform such other
duties as may be required of them by the general court or the Governor and Council.

SEC. 10. Whenever any highway or bridge, or public work appertaining to these two, shall come partly within the province of
this Board, and partly within that of any other State board, already
constituted, then such subject shall be discussed and decided upon
in a joint convention or conventions, composed of equal numbers
of this and the said other State board, and some member by them
chosen as presiding officer.

SEC. 11. All applications or propositions for improvements or
new works, of the kind specified in section nine as coming within
the province of this Board of Highways and Bridges, and intended
to be laid before the legislature, shall hereafter be first made to this

Board. Upon receiving such application, Board shall investigate same, and if they find such work necessary and proper, shall thus report to the legislature, with an estimate of the expense thereof; if they do not approve of such application, they shall report the reasons for their disapproval.

The Board may also, in like manner, recommend, whenever they think proper, any improvements of the kind above specified, though no application has been made therefor.

SEC. 12. It shall be the duty of the Board to procure for the legislature full plans and estimates of contemplated works or improvements when so ordered by the legislature.

SEC. 13. Whenever any work shall have been authorized or ordered by the general court and the money appropriated therefor, Board shall advertise for proposals for doing said work; plans and specifications of the same first to be placed on file in office of Board, which plans and specifications shall be open to public inspection; advertisement to state work to be done and to be published ten (10) days at least. The bids shall be sealed bids, directed to Board and accompanied by bond to the Commonwealth signed by bidder and two responsible sureties, in sum of two hundred ($200) dollars, conditioned he shall do the work if awarded to him, in case of his default to do so, forfeits, &c. Bids to be opened at time and place mentioned in advertisement.

SEC. 14. All contracts shall be awarded to the lowest responsible bidder and who sufficiently guarantees to do work under superintendence and to satisfaction of Board; provided that the contract price does not exceed the estimate or such other sum as shall be satisfactory to Board. Copies of contracts to be filed with state auditor.

SEC. 15. Board reserves right in contracts to decide questions as to proper performance of work and meaning of contracts; in case of improper construction may suspend work and relet the same; or order entire re-construction; or may relet to other contractors and settle for work done, &c. In cases where contractor properly does work, Board may in their discretion as work progresses, grant to said contractors estimates of amount already earned, reserving fifteen per cent. therefrom, which shall entitle holder to receive amount, all other conditions being satisfied.

SEC. 16. In case prosecution of any public work be suspended, or bid be deemed excessive, or bidders be not responsible, Board may, with written approval of governor, where the urgency of the case, or interests of the Commonwealth require it, employ workmen to perform or complete any work ordered by the legislature: provided, that the cost and expense shall in no case exceed the amount appropriated for the same.

SEC. 17. All supplies of materials, &c., when costing over five hundred ($500) dollars, to be purchased by contract, subject to same conditions as letting out work.

SEC. 18. Whenever Board think necessary, for interests of the Commonwealth, to protect same from damage or loss, shall report thus to governor and council and reasons for same, asking power to give contracts without notice required above, and governor and council may grant request, provided three-fourths vote for it.

SEC. 19. Whenever Board is of opinion a work may be done better without a contract, shall so report to legislature, and they shall procure machinery, materials, &c., hire workmen, &c., to do said work, whenever so authorized by the legislature.

SEC. 20. All contracts and bonds by Board to be in the name of the Commonwealth.

SEC. 21. No member of the Board to be interested in any contract; all contracts made with any member interested, governor may declare void, and shall remove such member so interested from office. It is the duty of every member of the Board and every officer of the Commonwealth to report any such delinquency, if discovered.

SEC. 22. Board shall be empowered to employ such engineers, clerks or other assistants, as shall be provided for by the legislature.

SEC. 23. Board shall, on or before   every year, submit to the auditor, by him to be presented to the legislature with his annual estimate, a statement of the repairs and new work needed for the current year, and of the sums required by the Board therefor; report to be in detail; all sums appropriated therefor to be included in the annual tax-levy.

SEC. 24. All moneys to be paid to any person out of moneys so raised, shall be certified by president of Board to auditor, who shall draw warrant on treasurer therefor, stating to whom payable and to what fund chargeable; such warrant to be countersigned by president of Board.

SEC. 25. Board to keep accounts, showing moneys received and spent, clearly and distinctly, and for what purpose. Accounts to be always open for inspection of auditor or any committee appointed by the legislature.

THE CONSTRUCTION AND MAINTENANCE OF ROADS.*

The writer wishes to give the Society some statistics and suggestions regarding the construction and maintenance of wheelways, partly drawn from his experience in charge of such work and from observation and information acquired in this country, and recently in London and Paris.

EARTH ROADS.

In the construction of an earth or gravel road the effort should be to keep the material near the surface as nearly homogeneous as possible, that the surface may be uniformly hard. The upper layers at least should be thoroughly rolled in thin layers, with sprinkling, if the material is too dry to pack well. The most efficacious roller for this purpose is of two sets of disks, one about eight inches less in diameter than the other, placed alternately on the axis of the roller. It is understood that the cost of compacting reservoir embankments with this roller is about three-fourths of a cent per cubic yard. The writer has never seen it used in road maintenance.

When the soil is sandy but little can be done besides covering it with some more tenacious material. Clay, or clay hard-pan, is the best, unless a sufficient coat of gravel can be afforded. Even a clay road, if the traffic is not too heavy, can be kept in a firm state by careful and continuous maintenance ; a coat of sand or hard-pan is, of course, desirable.

The plan often pursued of repairing roads once or twice a year is not economical, for the dilapidations increase in a heavy ratio after they commence.

A fair average for re-forming a mile of old road 30 feet wide between gutters, where the material was mostly cast from the sides, was 164 days, 10 hours each, of laborers, and two days of a double team hauling earth, carrying away stones, and moving tool-box. The use of a railroad scraper would have been an economy.

Maintenance is most economically performed by double teams, with hones or scrapers, rollers, and the watering-cart in dry weather. The hone or scraper is often an oak plank, 2 inches thick, 10 inches broad, and

*By Edward P. North, C. E., Member of the Society, read before the American Society of Civil Engineers, April 16th, 1879.

9 or 10 feet long, shod on its lower edge with a ¼-inch plate of iron, Drawn vertically along the road either by a tongue or a chain—in the first case it has two handles like plow handles—in the second it has a vertical handle, and the earth is dumped by pushing it forward; a piece of plank about 3 feet long being fastened behind, by riding on which the driver can regulate the amount of earth moved. These are drawn over the road, inclined 7 or 10 degrees from a perpendicular, to the line of travel. So that, besides filling small depressions, they, to a slight extent, scrape the earth to or from the centre of the road. There are also some patented machines, combining a scraper and roller in a frame, which are said to be very effective. 46,000 to 47,000 square yards can be covered in a day, while not more than 2,300 can be put in order if the road is rutted and gullied. While hones are of little use on muddy roads, they are effective just as it is drying.

The ordinary type of roller has 6 rings, the whole length being 6 feet, and weighing about 2 tons, with frames that will hold 1½ tons of stone. A better is of 2 rings, 3 feet long, also weighing 2 tons. Another is figured in Clemens Herschel's Prize Essay on Roads, and in General Gilmore's Roads, Streets and Pavements.

The roller should follow the re-forming of the roadbed, whether with hones, shovels, or the ordinary railroad scraper.

The writer has rolled earth roads with a 15-ton steam roller, but not enough to be certain as to its economic value; where the soil contained a fair amount of clay, the roadbed was left very hard, and wore well.

A water cart holding 60 cubic feet will water 830 to 860 square yards, and can be drawn by an ordinary team over any road that is worth watering. Two sprinklings per day will keep the road in good condition, though not free from dust in hot weather or high winds, if there is much traffic on the road. Sprinkling is the only thing that will keep a road from breaking up in long continued dry weather.

The treatment of gravel roads compacted either by traffic or horse rollers, differs very little from that pursued with earth roads. When the gravel is over 1 inch in diameter, it is almost impossible to keep the roadbed from breaking up when dry, and ¾ inch would be a better size, unless continuous watering can be depended on in dry weather. Small gravel (and the same remark applies to metal for Macadam) makes a pleasanter road for travel, and can be more easily kept in order.

St. Nicholas Avenue, which will be mentioned further on, was made from nearly clean Roa Hook gravel, by the aid of a 15-ton roller, but with a horse roller it will be necessary to add clay, loam or some softer material, to any hard or clean gravel to act as binding. W. H. Grant, Member of the Society, in his valuable description of the roads of Central Park, says of Roa Hook gravel, " it being more than ordinarily clean and hard, bears an intermixture or adulteration of 20 to 25 per cent of inferior material to perfect its binding properties." These roads have a foundation of rubble stone, not so firmly packed as Telford specified, covered with quarry chips and hard-pan, which was rolled with 2-horse rollers, 6 feet long. "This is thoroughly done to prevent the gravel filling the cavities of the rubble bottom, so that its cellular character may remain unimpaired to facilitate drainage." The gravel was applied to two or three successive layers, making a depth of 4 to 6 inches. Each layer was rolled with a 2-horse roller, and the last with one weighing 6½ tons, 5 feet long, giving 217 lbs. per inch pressure. These roads, which were thoroughly underdrained and side guttered, have long been famous for their excellence ; they are pleasanter for light travel than Macadam, and are easily kept in repair, except in wet weather, when they become muddy, and when neglected, the larger pebbles make a rough road for buggy riding.

For horse rolling binding should be applied as sparingly as possible, and on the last layer, after it has been compacted, simply as a glaze to hold the stone ; if it contains clay, it should be as moist as possible, not to stick to the roller.

The so-called Tompkins Cove gravel, which is much used for entrance drives to gentlemen's places about New York, is a broken limestone, apparently of the cement series. It is usually spread over the road, and compacted by the wheels. The darker colored stone is very pleasant to the eye, and it readily makes a smooth wheelway singularly free from either mud or dust, even when subjected to rather heavy traffic, though it is too friable for economical use in such situations. Its performance is so different from that of the ordinary limestones that an analysis is appended :

```
Lime ...........................................................  60.20
Alumina........................................................  11.22
Silica..........................................................   6.13
Magnesia.......................................................  10.45
Carbonic Acid..................................................   8.00
Water..........................................................   4.00
                                                                -------
                                                                100.00
```

MACADAM.

Roadways surfaced with broken stone have been in use for a long time, but Macadam, about the end of the last century, systematized their construction and maintenance, effecting such an improvement in English roads that his name seems indissolubly connected with roads of this class. His plan seems to have been essentially to have a well drained, uniform road bed for the reception of small, clean stone of uniform quality applied in thin courses, deprecating any addition of earth, clay, chalk, or other matter that will imbibe water and be affected by frost, under the pretense of binding it.

He held that a surface of an inch square or a stone 6 ounces in weight was of the maximum size for road metaling, anything larger than that being mischievous, and that 10 inches of solid Macadam was sufficient to carry any load, rather preferring a soft substratum, saying that the cost of maintenance on a morass was to the cost on a rocky foundation in the ratio of 5 to 7.

Telford differed from Macadam in his views, how much may be best seen from one of his specifications taken from Parnell's Treatise on Roads : London, 1833, p. 147 *et. seq.* : " Upon the level bed prepared for the road materials a bottom course or layer of stones is to be set by hand in the form of a close, firm pavement.   They are to be set on their broadest edges lengthwise across the road, and the breadth of the upper face is not to exceed 4 inches in any case. All of the irregularities of the upper part of the said pavement are to be broken off by the hammer, and all the interstices to be filled with stone chips firmly wedged or packed by hand, with a light hammer.   The middle 18 feet of pavement is to be coated with hard stone as nearly cubical as possible, broken to go through a 2½-inch ring, to the depth of 6 inches, 4 of these 6 inches to be first put on and worked by traffic, after which the remaining 2 inches can be put on. The work of setting the paving stones must be executed with the greatest care, and strictly according to the foregoing directions, or otherwise the stones will become loose, and in time may work up to the surface of the road. When the work is properly executed no stone can move : the whole of the materials to be covered by 1½ inches of good gravel, free from clay or earth."   Of which Parnell says, " The binding which is required to be laid on, on a new-made road, is by no means of use to the road, but, on the contrary, injurious to it. This binding by sinking between the stones diminishes the absolute solidity of the surface of the road, lets in water and frost, and contributes to prevent the complete consolidation of the mass of the broken stones."

As regards the foundation, Macadam seems to have the engineers of France and England mostly on his side. In this country, it is believed engineers generally prefer a Telford foundation. Neither Macadam or Telford used a roller, both depending on the grinding action of the wheels of wagons to compact their roads ; and it was probably to lessen the brutal pulling through loose metal that Telford used his coating of gravel.

Road rolling was first brought prominently before English readers by Sir John F. Burgoyne, in a paper written in 1843 (which is reproduced by Law & Clark).   And the best American practice very fully indorses his views. He recommends a weight, as the greatest attainable, of 261 pounds per inch run of roller, the use of stone dust or sharp gravel for binding and watering. His opening sentence, " the importance of rolling roads, either newly constructed or when subjected to extensive repairs, seems never to have been duly appreciated," is still true in England, for the writer saw loose stone kicking about Great George street, in front of the Institution of Civil Engineers, for over a month.

It may here be said that writers on roads have not always kept the differences between the possibilities of construction of the three types of Macadam roads (viz., traffic made, horse rolled and steam rolled) sufficiently before their readers. The grinding action of wheels will pack the hardest stones of proper size, and make a firmer roadbed than if a softer material is used for binding, but about one-third of the material is worn out in the operation, to say nothing of the wear on horses and wagons. On account of the digging action of the horses' feet when drawing heavy rollers, horse rollers will only compact and bind the softer rocks, without the aid of binding material, and no steam roller known is heavy enough to bind trap and the harder granites, without such aid.

The only circumstance that can justify an engineer in depending on traffic to make his road (except in the case of small, thin patches) is an inability to procure a roller ; and then the material should be applied in very thin layers, for, in addition to the waste of material, the cost of raking and leveling a new road will nearly equal the cost of horse rolling, the surface will never be as good ; and, as dung and dust will be ground in with the stone, it will be more affected by wet and frost. In addition, as a road is made for economy and convenience of transportation, the damage to horses and vehicles should not be allowed, even if it made a better road.

While a horse roller will make a much better road than traffic will, it is inferior to a road rolled by steam rollers, for about 260 pounds per inch run of roller seems to be the greatest weight practically attainable by horse rollers, which is not sufficient to make a road of trap and the harder stones, without the use of some softer materials as binding, which presents the objections stated by Parnell, even under the most careful system of application.

With a steam roller the weight applied can be made equal to the requirements. The horses' feet do not cut up the metal almost as fast as it is arranged and compacted by the roller. Hard binding can be used, making the roadbed nearly homogeneous and impervious to water, preventing movement of the stones on each other. Lastly, the frictional action of the driving wheels arranges and compacts the stones better than a greater rolling weight does, making, in fact, an entirely different wheelway in its wearing and sanitary aspects.

The best horse-rolled road known to the writer is the Southern Boulevard, built by Wm. E. Worthen, Member of the Society. The earth was compacted by rolling, on which 2½ inch trap was placed in one

6-inch and one 4-inch layer, when both layers were compressed as far as possible by the use of 2-horse rollers ; 2½ inches of screenings were spread over and rolled in. As water was not easily attainable on the line of the road, the screenings were thoroughly wet before they were carted to the road. This wheelway, which was fourteen or fifteen feet wide, stood seven years, almost entirely without care, under a heavy cart-ing traffic and a good deal of light driving. But the road was never as solid as those made with steam rollers and properly puddled, opening more with the frost, and having more loose stones on it.

The writer built some horse-rolled road on an old Macadam road which was substantially worn out. The material was a syenitic gneiss quite hard, and broken by hand to pass through a 2-inch ring. It was generally laid on 6 inches thick, in one course, though where the old metal was mostly gone, two layers of stone were used. The rollers were 6 feet long, weighing about 2 tons light and 3½ loaded. They were used with 2 horses light and 4 horses heavy. The stones, which were broken on the side of the road, were handled with 10-tined forks, the tine 14 inches long, 1¼ inches apart, known as "tanners' forks." These forks, with a little care on the part of the laborer, left nearly all the dust and small stone behind (the average wear of tines was one inch for 300 cubic yards handled), so that the stones went into the road fairly clean, where they were rolled ; the shoulders being at the same time made up with good earth, until they were fairly compact, when the dust and small stones left by the forks were spread on the surface, which was again covered with about ¾ of an inch of preferably clayey soil, where it was procurable, which was rolled thoroughly. If the binding was sufficiently damp, the road stood very well, showing few loose stones ; but when it was too dry or the earth was washed into the body of the metal by heavy rains, the work was not at all satisfactory.

The labor account per mile of wheelway, 14 feet wide, equal to 8,213 square yards, on which 1,260 cubic yards of broken stone were placed, was as follows, the days being 10 hours each : Spreading and forming material and loading dirt for shoulders and binding, with sweeping the old Macadam clean, but no picking, 229.5 days' labor. Twenty-four days were occupied rolling, 13.2 with 2-horse and 10,8 with 4-horse teams.

This road, with the shoulders, was 18 feet wide, and the teams trav-eled at the rate of 2 miles per hour ; allowing 90 per cent. of the time as productive, the roller passed 144 times over the surface. The mean weight of the rollers per inch run was only 75 lbs., and the maximum 100 lbs., a weight altogether too light for either economical or thorough work.

One end of this road was on a hill, with a rise of 110 in 1,500 feet ; the maximum grade being at a rate of 8 per 100. It was not judged advisable to depend on natural moisture for the binding, and a steam pump was procured, by the aid of which the metal was made thoroughly wet before the binding was applied. The binding used on the hill was a light loam ; but considering the grade, the finished wheelway was as good as when heavier loam had been used, relying on the moisture in it.

In England there does not seem to be any well-established system of road making. Some of the borough surveyors apparently do not believe in any rolling at all ; others, after carefnlly picking up the old roadbed, put on the stones, and after more or less traffic has passed over it, put on a 5-ton roller 5 feet long, drawn by 4 to 6 horses. The binding called " hoggin," which is a loam with coarse sand and gravel in it, is apparently not always applied in the case of repairs. Appendix No. 1 is referred to for specifications for a new road.

The repairs of the wheelway of the Victoria Embankment were made by first thoroughly picking up the old roadbed ard then coating it with clean Guernsey granite, hand broken, to go through a 2½ inch ring, entirely free from dust and debris, and nearly so from small stone. Guernsey granite is about as hard as a trap. It was rolled with 15-ton Aveling & Porter's rollers. When compacted, hoggin was added in sufficient quantity, and with enough water to flush the material into all the interstices and leave a surplus about the thickness of grout on the surface, which was swept with brooms in front of the rollers. The same system was employed at Russell Square and Bedford Place.

This did not make a good Macadam road ; the hoggin acted as a lubricant, allowing the stones to work on each other under the traffic. There were loose stones on the road ; five weeks after the road was completed, a kick would move the stones in front of it for fully a foot ; when the surface was drying, the outlines of many of the stones could be seen by the cracks in the mud covering them ; the angles of the stones were already wearing off, and after a rain, on sweeping the road, the gutters were filled with mud 4 to 6 feet wide.

It was said that dissatisfaction was felt with this system, and during the past summer an effort was made to make a road simply by rolling, without the aid of any binding. The success was no greater than that of Mr. Grant, 20 years ago, in the Central Park.

In Liverpool a mixture of broken stone and coal tar pitch is laid— one ton of pitch that will just run at 100° F., is mixed with 70 galls of dead oil, and added, hot, to 5 cubic yards of stone. The stone is *dry*, but not heated ; the mixture is rolled with a hand roller in short lengths. The surface is very good, making a favorite road for bicycle riders, and it can be swept without danger of dislodging stone, but it does not seem to have the wearing qualiiies of a well puddled trap road. Some laid three years in Basnat street, now requires pretty extensive repairs.

In Scotland, a concrete has been used with Portland cement, for binding; the surface was very good, but when the road commenced to break it went to pieces very fast.

M. Malo speaks of the Macadam in Paris as follows : " Day by day they try with incalculable efforts to perfect Macadam. They have perfected their watering in summer and sweeping in winter ; they have substituted, at great cost, granite in place of limestone ; they have multiplied their road laborers ; each morning the devastacions of the day before are repaired with incredible rapidity, but when the stream of traffic again covers the street, the scourge (*fleau*) recovers its rights, and during 300 days of the year the roadway becomes an ocean of mud or a mass of infected dust," In other words, a poor system of construction, but probably the most thorough system of maintenance to be found anywhere. A. Debauve, in his "Manuel de l'Ingénieur," after speaking of the immense capital invested in the old Roman roads, says : " The tendency now is to have the thickness only that which is necessary for resistance, to suppress the foundation, and, in one word, to economically establish the roadways and maintain them afterward in a fit state by incessant repairs, which the ancients did not know of."

In Paris the Macadam roads are composed of water-worn flint pebbles, which are compacted by ramming with a rammer 8 inches in diameter, weighing 70 lbs., and horse rolling with the aid of sand and water, *meuliere*—a kind of burr mill stone—and porphyry, the two latter are generally steam rolled. These materials are used, according to Debauve, in the following proportions, viz. : flint, 10 per cent. ; meulière, 67 per cent., and porphyry, 23 per cent.

Macadam roads in Paris, as in London, are the roads of luxury. On many streets, the centre, for a width of 19 feet, is covered with Macadam, while the sides, for a width of 13 feet each, are paved with stone sets or asphalte. The Avenue des Champs Elysées, Place Concorde, and some of the Quais are of flint ; other of the Quais and Rue de Rivoli are of harder stone, while the Avenue de l'Opera, Boulevard Haussmann, etc., have Macadam only in the middle.

The French, who use the Gellerat roller in Paris, specify a maximum weight of 448 lbs. per inch run, and a maximum speed of 2.3 miles per hour. The rolling is done by contract (the city furnishing the water) at a rate per ton mile, varying from 15.26 to 7.63 cents, according to the amount, with an increase of one-third in price where the grade exceeds 6 per cent.

The thickness of metal rolled varies from 12 inches on new roads to 2 inches on old. It is maintained that the work of compacting a cubic yard of the same stone is independent of the thickness when that varies between 2 and 6 inches, and is from 2.7 to 3.27 ton miles per cubic yard.

From a table (p. 308) in the notice of the objects, etc., exhibited by the city of Paris at the Exhibition of 1878, it is seen that the mean weight per inch run of the steam rollers is 448 and 336 lbs., and for the horse rollers 263 lbs. Also that the ton miles necessary to make a square yard of porphyry wheelway, or to compact a cubic yard of the same metal, are as follows : the mean for the two models of machines weighing 448 lbs. per inch run was, per square yard, with thickness of 3.9 inches, 0.41 ton miles ; while for the roller of 336 lbs., with a thickness of 2.8 inches, 0.234 ton miles were required, or 3.78 and 2.99 ton miles per cubic yard respectively ; and for horse rollers, where the thickness was 2.6 inches, the ton miles required were 0.194 per square yard and 2.69 per cubic yard. The amounts consolidated per ton per hour are in the following proportions : 467 for the heavy rollers, 539 for the light roller and 297 for the horse roller, and the number of passages of the rollers were 98.5, 75 and 92.

It appears from these tables that the smaller weight is the more advan_ tageous, and that horse rolling is cheaper than steam rolling ; it is added, that, on account of their superior celerity, the steam rollers are almost exclusively employed.

No statement is made as to the relative wear of stone rolled by the different machines.

About 24 per cent. of sand is used for binding. The surface is allowed to wear down until 2 to 6 inches of metal is put on.

Repairs were being made on Boulevard Haussmann by a 5-inch layer of meulière, very nicely hand broken to pass through a $2\frac{1}{4}$-inch ring, free from dirt and small stone ; the roller had a weight of 360 lbs. per inch. After a few passages had been made over the stone, very clean sand was spread on, which was shortly wet by a hand hose. The water was not at first applied in sufficient quantity to flush the sand thoroughly into the interstices, but the sand was just damp enough to pick up the stones. The speed of the roller was thought to be fully $2\frac{1}{4}$ miles per hour. The binding was not so thoroughly wet as on the Victoria Embankment, nor was the metal rolled as much. When the street was thrown open to traffic the consolidation was about at the point at which in this country we would commence applying screenings, stones were picked up by the wheels, and a light kick would move the stones for from 1 to $1\frac{1}{2}$ feet in front of it.

On the Avenue de l'Opera, the centre of which is covered with broken porphyry, the stones presented the same evidences of motion on each other, mentioned as seen on the Victoria Embankment.

In New York City there are, including St. Nicholas avenue, 27.99 miles or 921.400 square yards of steam-rolled Macadam roads, besides the horse-rolled roads in the Central Park and the Twenty-third and Twenty-fourth Wards, of which 17.75 miles or 718,200 square yards are on Telford foundations. They were commenced in 1869, and mostly finished in 1876.

The first had a 5-inch layer of gneiss laid on the Telford, on that a 5-inch course of trap, both broken to go through a 2½-inch ring. The trap was and is machine broken, and the screenings, *i. e.,* dust and stone that passes through a 1-inch or 1¼-inch aperture, are used for binding. At first it was held that these should be used as sparingly as possible, but experience has shown that the roads wear better and have less loose stones on them when the interstices are fully filled with screenings, and the size of the stone in the top course is preferably from 1 to 1½ inches. The general practice now is to put on the stones in two courses when the thickness, compacted, is 6 inches or more, and roll the first course until it will settle no more, then add the last course, and after it has settled, screenings are added, at first coarse and then fine, and after thorough rolling, the whole is puddled by a copious addition of water. This is accomplished more thoroughly and satisfactorily in hot, dry weather. After the road is puddled ¼ to ½ inch of screenings are spread over it, and after drying an hour or so the traffic is turned on.

For the old style of construction Sixth avenue, between One Hundred and Tenth street and Harlem River, may be taken as a type, having the materials and courses as above mentioned. It was rolled with Aveling & Porter's 15-ton rollers, of the old pattern, in which the weight is nearly equally divided between the steering and driving wheels, which cover 6 feet, giving a bearing weight of 467 lbs. per inch run. As near as can be ascertained, 24 6-10 hours, at 1½ miles per hour, were occupied in rolling 1,000 square yards, giving for the work done 0.553 ton miles per square yard, 2.246 ton miles per cubic yard, and 129.8 trips of the roller over the surface.

Avenue St. Nicholas, which, as mentioned above, was made from Roa

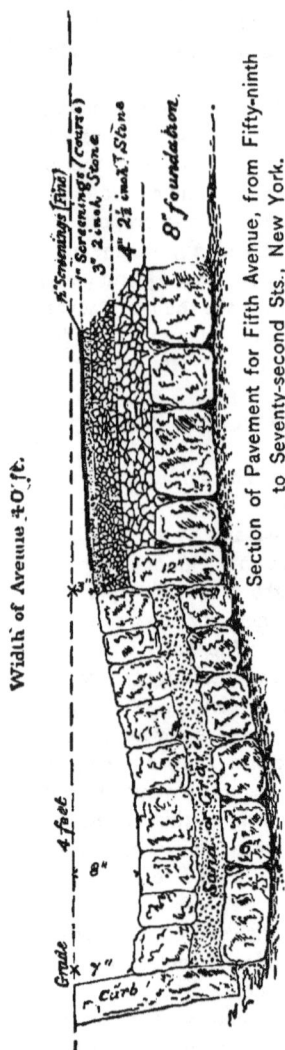

Hook gravel, had a rubble foundation, and it was intended to have 6 inches of compact gravel, but the foundation settled so that in some instances the gravel is 14 inches thick. The practice was to lay the gravel on and wet it thoroughly over night. It was rolled the next day until the water was forced to the surface. A 15-ton roller was employed about 26 hours per 1,000 square yards, besides some horse rolling. A little hard pan was used for binding. The intention was to make a road for fast driving, but the surface was as hard as that of a Macadam road, though much smoother.

The data for these two avenues depend on the memory of the engineer of the roller, and are approximate only.

The Telford Macadam on Fifth avenue has 8 inches of Telford, 3 inches of 2½-inch trap, 4 inches of 2-inch trap, 1 inch of coarse screenings, and one-half inch of fine screenings, when completed. The wheelway is 40 feet wide between curbs, with 4 foot gutters on each side, of trap blocks.

Mr. F. H. Hamlin, for some time Engineer in the charge of Roads and Streets in New York City, to whom I am indebted for these data, intended, by the use of the 1½ inches of screenings, to have the road as smooth and pleasant for riding as Avenue St. Nicholas, without the tendency to mud in wet weather. The work was let in two contracts, at $1.30 and $1.49 per square yard. The average time employed setting a square yard of Telford was 20½ minutes. Spreading the stone and screenings, per square yard, 1 2-10 minutes. About 58 6-10 hours rolling were given per 1,000 square yards ; and the work done, allowing the effective speed to have been 1¼ miles per hour, was 1.099 ton miles per square yard, and 4.709 ton miles per cubic yard of compacted material, if there had been no settlement, but there was an unascertained settlement.

These roads are maintained by spreading very thin layers of fine Roa Hook gravel over them and watering. The first roads were built with an idea that as small a quantity of screenings as possible should be used, and in dry weather the stones, which become prominent by wear, are liable to kick out ; the fine gravel prevents this, and by retaining water sprinkling is not necessary so often. The fact that the travel concentrated near the gutters, where this gravel, by the action of rain, becomes thickest, in connection with the demands of owners of fast horses, who wished a softer road-bed, led Mr. Hamlin to cover one of them with a coating three-eighths of an inch thick, consisting of three parts of coarse sand and one part of strong clay. This has been on the road through two winters and the road-bed is still firm, and it has been applied to others.

Under Mr. Hamlin's charge the average cost of maintaining these roads has been 4 2-10 cents per square yard per annum for the labor and material of road covering, cleaning and sprinkling, with incidental repairs and supervision. None of them have been resurfaced with broken stone excepting a portion of the Western Boulevard, the first built, and two or three patches of 100 feet or less in length, which cost 46 cents per square yard.

It is very difficult to estimate the traffic on these roads. On a fine day 3,000 vehicles may pass over them, while at other times there may be not more than one-tenth of that number.

In repairing the Southern Boulevard, mentioned above, the trap broken to go through a 2-inch ring was laid on 6 inches thick in one course; 38 2-10 hours' rolling was given per 1,000 square yards. Allowing the speed to have been 1½ miles per hour, the work done on it amounted to 0.859 ton miles per square yard and 5.177 ton miles per cubic yard; 201 trips were made over the surface. The Macadam was 15 feet wide, and some of the rolling was on the shoulders, though probably not enough to affect the result materially. The work was done in July and August, and a little less than 0.6 cubic feet of water per square yard (3.5 cubic feet per cubic yard) were used in compacting and puddling. About one-third screenings were added. Those portions of the road that had the most work done on them are now in the best form, after nearly two years without any care; and, generally, the more thoroughly the trap roads were rolled the better do they wear.

It should be understood that there was no counter on the wheels of the rollers, and the speed is the result of estimation, interfering with accuracy in the estimate of the number of ton miles performed. The rollers used in the above mentioned works were 15-ton Aveling & Porter's (old pattern), and though they are understood to bear equally on the driving and steering wheels, the writer thought the driving wheels nearly twice as effective as the steering wheels.

The superiority of the American Macadam roads is partly due to the greater amount of work put upon them; the binding, which is of the same hard material as the metaling, almost completely fills the interstices between the stones, and the entire mass is thoroughly compacted and nearly homogeneous. It is only while the frost is coming out of the road in the spring that the stones wear upon one another at all. The process of puddling gives a very smooth, hard, firm surface, resistant alike to wear and the infiltration of water, which is of equal advantage to the stability of the foundation and from a sanitary point of view. By the use of the steam roller the stones are compacted with a small amount of wear to their angles and an entire freedom from mud and dung, their only weak point being that in long-continued dry weather the larger stones are apt to get dislodged from the surface, some moisture being necessary for the full cohesion of the binding. No amount of wet weather, unaccompanied by frost, seems to injure them, unless mud works up through the foundation, and their imperviousness under the most disadvantageous circumstances is well illustrated by the practice of Mr. Hamlin, as cited above.

The English roads are rolled less, and the binding, though cheaper in cost and the matter of rolling and spreading, is, when thoroughly wet, to a great extent an element of weakness to the roads, allowing a large part of the wear to be internal, and failing to hold the stones from dislodgment, either by the wheels of vehicles or the brooms of the sweepers. When dry it can have neither cohesion nor resistance, its use being confined to the time it is slightly damp.

The roads in Paris seemed to be less thoroughly compacted than in London, but the binding was better, and if it had been applied dry, after the stone was nearly compacted, and only wet at the last rolling, it would probably be more effective. As it was, the grain of the sand seemed to have been broken down during the rolling of the stone, and as, like the English made roads, the stones move on one another, it must be still more thoroughly ground up.

The maintenance of roads in Paris is much more effective than in London. It consists essentially in washing rather than sprinkling them, and sweeping the mud and debris of the surface into the gutters, where it is washed with a copious flow of water, the mud and fine sand going into the sewers, the coarser sand being retained for use on the streets. Most of the work is done at night and in the early hours of the morning. While there is on the average (judging from the amount of mud relative to the traffic) rather more wear than in London, no loose stones were seen on the streets, except when newly mended, men apparently being always present with brooms, rammers, and sand, to repair any place that showed signs of weakness. On the Avenue des Champ Elysées, and other flint roads, the surface was often swept with birch brooms having long and slender twigs.

In London the maintenance is neither so constant or skillful; the roads are watered in dry weather, and swept or scraped in wet, with an occasional addition of sand or fine gravel.

There seems to be no accessible accounts of the cost of repairs and maintenance in London. W. S. Till, the Borough Surveyor of Birmingham, in one of his reports, says: " Mr. Howell, Surveyor to the Vestry of St. James, Westminster; informs me that the cost of maintaining, etc. the surface of Regent street, London, which may be considered one of the best Macadamized roads in the Kingdom, and in which nearly every description of pavement has been tried, is $0.87 per square yard per annum, it has, however, been much higher." It is not known whether Regent street was then steam or horse rolled.

In Paris, the expense of maintenance had reached on some of the streets 16 francs per square meter ($2.57 per square yard) per annum, and it was resolved to pave all but the middle 23 feet, in streets 48 feet wide, and to pave narrow streets and gutters, and in 1871 it was resolved to pave streets where the annual expense of maintenance was over 48 cents per square yard, excepting those avenues which are used by carriages of luxury. This, however, was not fully carried out.

Many adverse criticisms are made on Macadam roads by parties who judge of their value mainly from reading statements regarding European Macadam. A well-made trap road, when properly watered and maintained, is, after an earth road, the pleasantest and safest road known. In this locality a road 15 feet wide would give suburban residents the same easy access to their railroad stations in the worst winter weather as in summer. And for streets of residence, where the inhabitants would keep them free from garbage, both for quiet, safety to horses, and on sanitary grounds, they are preferable to the ordinary paving.

It is possible that with thoroughly compacted and bound roads, the English and French engineers would look with more favor on Telford foundations, for with the roads we make, the wear between the Macadam and Telford must be very slight, the pressure of a wheel being spread over a large surface. They are preferred here because they give an equa depth of foundation more cheaply than with broken stone, and as in this country municipal appointments of engineers are sometimes influenced by political considerations, the general desire is to build as solidly as possible.

One objection that may be urged against the Telford foundation is the fact that it does not have a solid bearing on the earth roadbed, and when the road is worn thin the spaces between the stones may become filled with water and mud, which will work through the foundation into the broken stone, hurrying the disintegration of the road.

The English engineers often use "hard core," a mixture of brick rubbish, old plastering and broken stone, on a clay soil, to prevent the mud working into the metaling.

The result was accomplished by the use of six to eight inches of fine sand on Mott avenue, New York, which was built at the joint expense of the city and private parties. The regulation of the surface required a maximum cut of two and a half feet and a fill of four feet ; the soil was a heavy loam, thoroughly saturated by the fall rains, which continued until the work was completed, and it was desirable to Macadamize it immediately. Rough gneiss rock was placed on the layer of sand and rolled to an even surface, with a thickness of about eight inches, and on that a scant six inches of loose 2-inch trap, to which thirty per cent. of screenings were added. The bottom was very treacherous ; on about half the road it was not safe to stop the roller, which broke through in two instances in spite of the care exercised. Nor could the road be rolled long in one place ; but as soon as the surface began to weave in front of the roller it was given a rest, and the roller taken to another part of the road. In spite of this the road was compacted and the surface puddled satisfactorily, except in one or two small patches. It has stood the freezing and thawing of two winters without receiving or requiring any attention.

It is held that the success of this operation, so contrary to the teachings of the books, was due to the layer of fine sand being impervious to the mud, which without it would have been over the top of the Macadam a long time before it was puddled.

The deductions of the French engineers from the table cited on page 12 should have been strengthened or modified by a statement of the wear of the different work done. And their principles of road construction are dissented from, because the writer believes that the horse-roller made road does not differ so much (after both are compacted) in wearing value from the road specified by Macadam as it does from the road properly made by a steam roller; that all binding should be fully as hard as the stone, and that the better road is made with the heavier roller per inch run, as far as his experience has gone, *i. e.*, up to 460 pounds. And, in addition, up to that weight the ton miles rather than the load should be altered on account of the greater or less hardness of the rock employed for road construction.

It is believed that no one who has used the two styles of 2-ton rollers mentioned, viz., 3 feet and 6 feet long, doubts that the 3-foot roller will do better work than can be accomplished by the 6-foot roller without loading.

Under the writer's direction the same quality of syenitic gneiss mentioned on page 9 was rolled with a 15-ton steam roller, with binding of the same quality in both cases. There were only about 200 cubic yards of it, and on account of the traffic passing, it was impossible to keep an exact account of the work done, but the road is very satisfactory, and from it the writer thinks that a few passages of the steam roller over the horse-rolled road would add materially to its life, besides greatly reducing the number of loose stones on its surface.

Some refuse Westchester marble (a very soft rock) was delivered on some of the roads at about 25 cents per cubic yard, and hand broken in place at the rate of about 1 cubic yard per hour. A portion was rolled before any traffic went over it, some after about two weeks of traffic and some after six weeks; of the rest, part was horse rolled and part compacted by wheels; the quality of the stone, traffic, etc., were very nearly the same; that not rolled by the steam roller soon wore into holes; the first mentioned is, after standing two winters, in very fair surface; the others decreasing in the order in which they are mentioned. This difference is so noticeable that any one could pick out their sequence as mentioned.

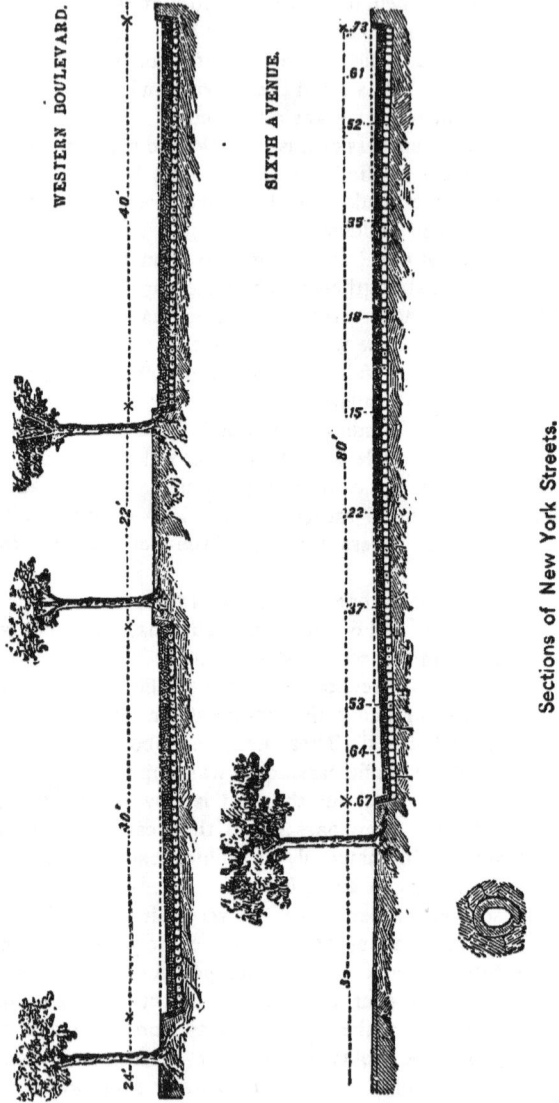

WESTERN BOULEVARD.

SIXTH AVENUE.

Sections of New York Streets.

Shortly before this, however, the writer very nearly made a mush of some micaceous gneiss in trying to reduce the crown of a road covered with it.

Late in the fall a portion of the Southern Boulevard was repaired with 2-inch trap, screenings, and a 2-ton horse roller. After the metal was compacted a thin coat of clay hard-pan was added, which froze solidly that night, and a day or two after was covered with snow which stayed on till spring, thus giving it the most favorable surroundings possible. In the summer there was but little difference between it and the steam rolled part; but by fall it had commenced to deteriorate, and now its surface is but little better than that made seven years before.

Both the English and French prefer hand-broken stones for Macadam. An experienced breaker will make better shaped stones than any crusher can, and hand-breaking would afford employment to labor. The hand-broken stones mentioned above were delivered by contractors at $2.00 per cubic yard; the men who broke receiving on an average 82½ cents per yard. One man was thought to have broken at the rate of 4.5 cubic yards per day; about 6 to average 2½ yards or more. They stood, using hammers weighing 1¾ to 2 pounds, on very flexible handles. The average of all who worked could not have been much over 1 yard per day.

On the other hand, there is not enough fine stone for binding, what there is is full of dirt, and machine breaking is cheaper; the cost of crushing trap with a Blake crusher, after it is sledged, being understood to be less than 70 cents.

The crown or transverse section of a road should depend on so many different considerations that no general rule can be drawn for it. With Macadam made from hard material, the less crown, on many accounts, the better, as the surface is benefited by being kept damp. In narrow wheelways, particularly, an excessive crown throws all the traffic, as much as possible, in one line, whatever the material may be. On the other hand, in curbed streets, if the road is too flat, the heavy traffic tends to concentrate near the gutters. The crowns given to dirt roads are intended to drain off the water, but the longitudinal ruts soon defeat that object.

Transverse profiles of the Western Boulevard and Sixth Avenue, New York City, are shown on page 80. On the former, though the transverse slope is so slight, it is perceptible to a person riding in a vehicle with longitudinal seats, but is not noticeable when on a transverse seat. The cross section of Sixth Avenue shows the surface as the road was constructed by M. A. Kellogg, Engineer, in 1872 and 1873, with the crown at the same height as the top of the curb stones, and the figures above the dotted line joining the curbs show the average distance of the surface below street line, as found by careful levels taken this spring between 123d and 124th streets, a point where no broken stone had been added, the wear at the crown being 0.15 foot, or about ⅕-inch per year. The road is not too flat, as it now stands, to shed the rainfall.

The amount of water the writer found necessary to keep earth or Macadam roads from becoming dusty, was, for a well maintained earth road, at the rate of 71.3 cubic feet per 1,000 square yards, applied twice in a day, or say 143 cubic feet per day. In very hot or breezy weather this was not quite enough.

On the Telford roads of this city, 25 cubic feet, applied four times a day, are necessary per 1,000 square yards, say 100 cubic feet per day. One water cart, holding 79 cubic feet, waters 35,000 square yards four times a day, keeping it free from dust, except during windy weather.

In Paris about 27 per cent. of the surface is watered by hand hose. These are made of iron pipes about 6½ feet long, each end supported on castors and connected with leather or rubber couplings; the working end being a piece of rubber hose. Its cost is one-half that of watering by carts holding 46 cubic feet.

Stone Pavements.

The City of New York is largely paved with the Belgian pavement, *i. e.*, truncated pyramids of trap set in coarse sand. The sand soon becomes saturated with the water and mud of the streets, and the blocks working under the traffic become rounded. It is impossible to keep such pavements in fair surface, and they are as bad sanitarily as pavements can be.

The pavements now laid are of granite or trap blocks, 4 inches wide, 6 inches deep, and 8 to 12 inches long. They have parallel sides, and are laid in sand, probably forming the most efficient pavement ($1.91 per square yard), to be found anywhere. The open joints filled only with sand are objectionable, however, as forming receptacles for street mud and water.

The Guidet pavement laid in Broadway, with a foundation of concrete, has lasted very well, its increased wear probably compensating for the additional cost of the concrete foundation.

In Paris, the new pavements are of Gris, a hard sandstone, which is neatly dressed in blocks, 4 to 5 inches square, and laid either in sand or mortar, and porphry, about 4 inches wide, 6 inches deep, and 8 inches long, generally laid in mortar.

The best granite pavements are found in England. In London, where the soil is clay, the usual practice is to lay a foundation course of " hard core " which is well compacted by rolling, &c., and the stones are set on this in two inches of sand. The better class of pavements, on clay, have 3 to 6 inches of hard core, 9 inches of concrete, and two inches of sand. Col. Haywood, where soil is sand or gravel, puts down 9 to 18 inches of broken stone, or 9 inches of concrete, saying that there is little difference in the stability of the two foundations, but the concrete is apt to be replaced in a more satisfactory condition when the street has been opened.

In all the better class of London pavements, the sets are 3 inches wide, neatly split out with parallel sides, and set in blue lias lime grout, Aberdeen granite being chosen in preference to harder varieties, as it wears rough. Though Col. Haywood determined that the duration of the same stone varied directly with their size, these narrow sets are preferred, as making a smoother road, affording a better foothold for horses, and increasing the quietness and general comfort. The practice is to take up the sets when their surfaces become rounded, and redress them, after which they are relaid in streets of secondary traffic. The spalls are useful for foundations and Macadam. Col. Haywood estimates the total life of such paving stones to be 29 years.

In Liverpool, Geo. F. Deacon prefers the Welch traps and granites, which are rather harder than our trap. The pavement of North John Street, which has a traffic of 4,000 vehicles, averaging 3 tons each, per day, is of trap blocks, neatly split out, with parallel sides, 6 inches deep, and 3¼ x3¼ inches on the face, set on 10 inches of concrete; the joints are filled with gravel, about the size of a pea, free from sand, and then run with coal tar pitch. It had been paved very close, and the sets were so firmly held in place that there was hardly any rounding of the surface last winter, though it was laid in 1872. The surface was admirable, and showed very slight wear, and on sanitary grounds it is probably unequaled by any other pavement, except compressed asphalte, for there can be no percolation of the surface water. The only signs of failure shown by this pavement is where the boiler of the " Montana," weighing with its 4-wheeled truck 59 tons, passed over it, crushing some of the sets, and showing that stones 3¼ inches square on the face, are not large enough to stand loads of 15 tons per wheel.

Mr. Deacon thinks that the 3¼ inch square blocks present too many longitudinal joints, and now employs sets 3 inches thick by 6½ inches deep, and 5 to 7 inches long; the specification for thickness is that any 4 of them, when set side by side, shall measure 12 inches, and not more than 14 inches. It is said the specification is filled without extra cost.

WOOD PAVEMENTS.

In London, the principal wood pavements are Cary's, the improved Wood, Henson's, and the Asphaltic Wood. Cary's is laid on sand or gravel, a firm foundation being first made. The patent is for dovetailing the blocks on their ends, which is held by the inventor to give the pavement greater stability. The joints are flushed with blue lias lime grout. The advantage claimed for this pavement is the ease with which it can be repaired, the method being apparently to take out the defective block or blocks, cut off the broomed portion that overhangs the sides of the blocks, put in enough gravel to restore the surface, replace the block with the other face up, and run in the grout. A portion of Cannon street is laid with this pavement. It is said by persons on the street to require frequent repairs. Its power of resisting wear must depend on its foundation, and it can hardly prevent water from working into and through the foundation.

The " Improved Wood " Company started to make an improvement on the Nicholson patent, using two tarred boards, resting on sand, under the blocks. Ludgate Hill, and some other streets, were laid on this principle, but, under the heavy traffic, the two boards acted like a pump, pumping the sand up through the pavement. After trying and discarding one board laid on concrete, they now lay a foundation of concrete 6 inches thick, on which the blocks are placed ; the joints are filled with pitch and gravel. This modification apparently makes a good pavement, as seen on Ludgate Hill, where the first pavement has been renewed, and at St. Paul's Church-yard. No company now lays boards as a part of its foundation.

Under the name of Ligneo Mineral Pavement, a company laid some pavement of hard wood, but it proved so slippery that work under that patent has been mostly abandoned. The most work seems to be done, at present, by the Henson and Asphaltic Wood companies.

In the Henson pavement the 6 inches of concrete is covered with a layer of tarred felt paper, on which the blocks rest, while a strip of the same material is placed vertically between each row of blocks across the street. Nothing is interposed between the ends of the blocks. After each few courses are laid the blocks are driven close together by a heavy maul. Hot coal tar and pitch is plentifully applied to the upper surface, after which gravel is strewn over it. It will be observed that this pavement presents about as small a joint for foothold as is possible, and to meet objections on this score, at first about every fifth course had a V-shaped groove cut in its top. This has been abandoned by the company, who now lay the surface flush, making a road almost free from noise and nearly as smooth as a compressed asphalte pavement. In answer to objections to their narrow joints, their engineer asserted that they had changed the practice of the other companies in London from 1¼ or 1 inch joints to as near a quarter of an inch as it is practicable to make them, to the decided advantage of all pavements. They also assert that the tarred felt under the blocks will not wear out, that the pavement is impermeable to water, and that it can be more thoroughly cleaned than any other wood pavement. This pavement was taken up, for pipe connections, in Leadenhall Street and High Holborn. Although it had been raining in the first instance, and raining and freezing in the other, the wood was bright and not at all watersoaked ; the felt underneath, however, was considerably worn. The claim in regard to cleaning is probably correct.

The asphaltic wood pavement, laid under the patent of H. S. Copland, C. E., has on the 6 inches of concrete a quarter of an inch of tempered Trinidad bitumen, on which the blocks are placed; the joints across the street, which are kept between a quarter and an eighth of an inch, are run with a softer bitumen for about 2 inches, and flushed with lias lime grout, the ends of the blocks abutting against each other; the whole is covered with gravel. The pavement, unless cracked by the settlement of the foundation, must be as impervious to water as one of asphalte, and should stand the traffic very well. On account of its rigid foundation it is more noisy than the Henson, and having a slightly wider transverse joint, it hardly rides as smoothly after some wear, though the difference would not be readily noticed. It should be noticed that the bitumen which is run in the joints makes a firm bond with the layer under the blocks, and more wood is spoiled in taking up this pavement than any other observed.

This pavement was first laid in Cannon street, between Abchurch and Laurence Pountney Lanes, in July, 1874. It was laid, apparently, with joints rather over than under 1 inch, and in December, 1878, the edges of each block had broomed over into the joint; otherwise the surface was good, and I could not learn from any source that repairs had been made on it, except when the street had been opened for pipes. The grade is 1 in 90; and in 1873 the number of horses passing over it in 12 hours—from 8 A. M. to 8 P. M.—was 5,350.

The practice in London is to cover the surface of wood pavements several times a year with hard gravel, which is beaten into the ends of the fibres by the traffic, tending to preserve the blocks from wear and, it is claimed, from decay. The wood used is a species of pine—Baltic fir—harder than our white pine and softer than Southern or pitch pine, resembling what is sold in the Chicago markets as Norway pine; it is better seasoned than the pine generally used by house carpenters in this country, and it is usually laid without being creosoted, the borough surveyors claiming that the difficulties of inspection are increased by creosoting; but as far as noticed, for renewals and repairs, which are made by the companies under their contracts for maintenance, creosoted wood is used, their managers saying that they expect it to add to the wearing qualities of the wood as well as to protect it from decay. The concrete, usually covered with a thin coating of cement mortar, is made of one part of Portland cement to six parts of Thames ballast, which varies in size, from sand to pebbles three-quarters of an inch in diameter.

It is rather difficult to arrive at the durability of either wood or asphalte pavements in London, as Col. Haywood, the City Engineer, lets them at a certain price, with a provision that they shall be kept in repair for a term of years at a certain price per square yard per annum, a good pavement being turned over to the city at the expiration of the contract for maintenance, the contractor, in the meantime, replacing the entire roadway, if necessary, as will be noticed under the head of asphalte. This plan has worked so well in the city that the surveyors for the vestries have gener-ally adopted it, so that, in London, the wood paving contracts are usually let with the same length of maintenance as for compressed asphalte, and at about the same price, viz., at a first cost of from $4.38 to $3.90 per square yard, including foundation, but not excavation. They are kept in repair for two years at the contractor's cost, and at the rate, depending on the amount of traffic, of from 36½ to 18¼ cents per square yard per annum for the succeeding fifteen years.

From the report of Col. Haywood to the Commissioners of Sewers of the City of London upon asphalte and wood pavements, 1874, page 38, *et seq.*, we find that the greatest duration of any wood pavement was in Mincing Lane, nineteen years and one month ; the average cost per square yard per annum having been, in that instance, thirty-five and one-half cents, while the average cost of three heavy traffic streets had been sixty-three and nine-tenths cents. He concludes that, with necessary repairs wood pavements will last ten years in London.

Law & Clark's Roads and Streets, page 239, gives the wear of three wood pavements at three-tenths of an inch per annum under a traffic averaging 362 vehicles per day of twelve hours for each foot in width of the street. Mr. Clark says it is claimed, in some instances, blocks have worn down in London to a depth of two and one-half inches (the Wells street pavement was only two inches thick, before removal), and suggests eight or nine inches as a better depth than six inches, now universal. The blocks of the Mincing Lane pavement, which lasted nineteen years, were nine inches deep. If we assume that the blocks are six inches deep, and that the road will not break up until they have worn down three inches, there seems no reason why a thoroughly creosoted wood pavement, laid with narrow joints, on a sufficient bed of concrete, with a water-tight stratum interposed, should not wear for about ten years in our streets of heaviest traffic with a small amount of intelligent maintenance.

There have been several charges made against wood pavements, which are mentioned here, not as a matter of information, viz., that they soon become full of holes, are impossible to clean, are difficult to replace when the street is opened, and that by their rotting the health of the community is endangered, to which may be added that for only a short time do they present any barrier to the saturation of the soil by surface water.

The general practice, as far as observed by the writer, in this country has been to lay *green* or wet pine blocks, more or less thoroughly dipped in coal tar, on a bed of sand, not always thoroughly rammed, with or without the interposition of a tarred pine board, with transverse joints from one to one and one-half inches wide filled with gravel and coal tar, which was theoretically thoroughly compacted. Omitting those which, without the slight pretense of a 1-inch board for foundation, speedily become a wood Macadam, the first failure was from the blocks rotting on the bottom in patches, so that the surface would first be found to be *sheared* down by a heavy load; and on taking out the blocks they would be found sound on their tops, where the gravel had been driven into them by the traffic, and also a sound film of about the thickness that tar could be expected to penetrate a wet block, the inside being rotten. In other cases, however, when the layer of sand was too thin, the mud worked up through both boards and blocks, reducing everything to about the state of those having no boards under them.

A layer of concrete, covered with Trinidad bitumen, will effectually stop the mud from coming up, and any percolation of the surface water into the soil through the pavement; the narrow joints, by preventing the edges of the blocks burring over, will both tend to keep the surface smooth, lessening the shocks of the wheels, and greatly facilitate all the operations of cleansing. Creosoting, by the preservative and antiseptic properties of the dead oils used in that process, will probably keep the timber from decay, so that nothing but abrasion need be feared, and the sanitary objections to decaying wood will be removed.

The following note of some experiments by E. R. ANDREWS (published in " Engineering News ") shows the efficacy of creosoting for protecting wood from moisture.

The following are the results of some careful experiments with different varieties of wood, half of the specimens being simply dried and the others creosoted, to ascertain to what extent wood is rendered water-proof by creosoting. The specimens were soaked during two days in water, being carefully weighed before and after soaking :

	Percentage of water absorbed.
Spruce, creosoted	.0236
" "	.0306
" dried only	.1754
" "	.8333
" Burnettized	.2500
Hard pine, creosoted	.0000
" dried only	.1600
Oak, creosoted	.0625
" dried only	.2000
White birch, creosoted	.1240
" dried only	.4300
Cottonwood, creosoted	.3470
" dried only	.7140
Black gum, creosoted	.1250
" dried only	1.0000
Sesquoia Gigantea (great tree of California), creosoted	.0000
" " " dried only	.4722

In the rooms of this Society are creosoted fir ties from England that have been in the track for 20 years, and apparently justify the assertion of the engineer sending them, that they are good for 10 years more.

The following extract, from the " Railroad Gazette," is also corroborative :

The German Railroad Union some time ago made inquiries as to the extent to which processes for preserving ties were employed, and what the results were.

It appears from statistics of German railroads which have used treated ties more or less since 1840, and therefore have had time to test thoroughly the life of the ties, that the average life of ties not treated, and of those treated with chloride of zinc or creosote has been :

	Not treated.	Treated.
Oak ties	13.6 years.	19.5 years.
Fir ties	7.2 "	14 to 16 "
Pine ties	5.0 "	8 to 10 "
Beech ties	3.0 "	15 to 18 "

The average life of 831,341 pine ties treated in various ways on thirteen German roads was 14 years.

It follows from this that there is an increase in the life of ties treated with chloride of zinc or creosote, amounting to about 40 per cent. for oak, 100 to 130 per cent. for fir, 60 to 100 per cent. for pine, and 400 to 500 per cent. for beech.

It thus appears that there is a great deal gained with any kind of wood, but most with some of those usually not considered good for ties, fir and beech being made almost as durable as oak. Bischoff says that it is of little advantage to secure the ties from decay for longer periods than above stated, as the ties usually become worn out or crushed by that time, even if not decayed.

Commenting on these facts, Bischoff says that it is now generally admitted that the choice lies between creosote and chloride of zinc ; that creosote is the best antiseptic material, but also that it is the dearest.

There can be but little doubt that the antiseptic properties of the creosoting process are of more value than the increased life it would give to the blocks.

On account of the absence of proper stone and the cheapness of wood in large areas of our country, the small first cost of wood pavements seems to make it worth while to give them an intelligent trial.

The thoroughness with which wood pavements can be cleansed depends on the size of the joints and the firmness of their filling. The practice in London, when the mud is at all sticky and not so thick as to require scraping, is to water and then sweep with a revolving broom, the thoroughness of the cleansing being almost directly as the amount of water. In hot weather a disinfectant is sometimes applied after sweeping.

Asphalte and Bitumen.

Dictionary and encyclopedia makers, as well as chemists, seem to use these terms interchangeably. M. Leon Malo, in his " *Guide Pratique pour La Fabrication et L'Application de L'Asphalte et des Bitumes,*" after speaking of the lack of definition in the two terms as generally used, gives the following, viz.:

"*Asphalte*, Bituminous Limestone."

"*Bitumen*, The black and viscid substance which we find disseminated in the pores of bituminous limestone and in the interstices of the molasses of Seyssel or the sands of Auvergne."

"*Asphaltic Mastic*, Bituminous limestone transformed by dressing (*cuisson*) and by the addition of a small quantity of bitumen."

He adds: "We employ these definitions throughout this work, and give to them the senses which we have just indicated, and I hope strongly that they will be adopted for use, for they seem to be the most reasonable."

It is believed his definitions have been generally accepted by European technical writers and in specifications relating to the use of these materials,* and the writer in this paper will use them as above defined, excepting that the bitumens from Trinidad and Cuba will be included with those mentioned by M Malo.

In view of the practice in this country two other definitions seem necessary, viz.:

Bituminous mastics, mixtures of bitumen, either having an earthy *gangue*, like those of Trinidad and Cuba, or purer ones, like Grahamite and Albertite, with limestone or other substances not naturally impregnated, which add to its resistance to wear, and—

Tar Mastic, a mixture in which the bitumen is replaced by (usually) gas tar.

M. Malo further characterizes asphalte as a carbonate of lime perfectly pure (excepting, sometimes, a trace of silica), naturally impregnated with bitumen. Its characteristic color is that of chocolate—which it also resembles in fracture—mean specific gravity 2.235. It is quite hard when cold, and falls to pieces at a heat of about 122 deg.-140 deg. F. At an intermediate temperature it flattens under the blows of a hammer ; its structure varies with its locality, but in general should be fine-grained and homogeneous, without particles of unimpregnated stone.

In the poorer qualities the impregnation, though regular, does not exceed 6 per cent., or the bitumen is injected into minute cracks, showing under the microscope that the impregnation is not molecular, or the rock contains *clay*, or, as in the case of Lobsan, the bitumen contains light oils which injure the consistence of the mastic, in which case the light oils are driven off by heat, the remainder being used.

Generally, we may say that the more uniform and microscopic the impregnation of the lime is, and freer from extraneous matters, the better it is.

---

*In Paris, the asphalted sidewalks are universally spoken of as bitumen, in distinction from wheelways, which are called asphalte.

Neglecting the Tubal Cain stage of its history, asphalte was first applied to sidewalks and wheelways as a mastic, but in that state was not sufficiently resistant for streets of much traffic. It was seen, however, that in summer the pieces of rock that fell from the carts between the mine and the mastic works at Seyssel compressed under the wheels. In 1849, a Swiss Engineer, M. Mérian, put this lesson to profit, by constructing a Macadam road of asphalte, which was compacted with a roller. In spite of the instability of its foundation and the irregularity of its maintenance, this road is still in very good order. (Malo, p. 108.)

COMPRESSED ASPHALTE.

The first *compressed asphalte* was laid in Paris by M. Vaudry in 1854, though it was not till 1858 that it was laid on a large scale; the area covered in 1878 by pavements and cross-walks was 324,654 square yards, or nearly 14 miles of a street with a 40-foot wheelway. The earlier practice was to " decrepitate " the rock (broken to about the proper size for Macadam) by heating it to about 140 degs. F. in shallow iron pans. Skill was necessary in this operation, as the rock was liable to be burned, *i. e.*, have too much of the bitumen driven off, or to have too little, and sometimes both results were reached in the same batch. The walks in Union Square, N. Y., are an example of the ill effects of unskillful decrepitation.

The rock was also broken up by the direct action of steam, but it was impossible to secure a product of uniform fineness, and difficult to remove all the moisture. After experiments with various machines, a Blake crusher is first used, the pieces are then passed through a Carr's disintegrator, after which the powdered rock is heated in revolving cylinders to from 212 degs. to 284 degs. F. (depending partly on the distance it has to be carried), and transported to the place where it is to be laid, usually in covered sheet-iron wagons.

The permanence of this pavement depends primarily on the stability of its foundation, which is usually of concrete (though old asphalte is sometimes used) 6 inches thick; on this, when it is thoroughly dried, the heated powder is spread, by means of rakes, to such thickness that when compressed it shall be from $1\frac{1}{2}$ to $2\frac{1}{2}$ inches thick—depending on the probable amount of traffic. The compression, which is termed pilonnage, is effected by the aid of the tools figured. The fouloir is first used along the junction of the asphalte with either the curb-stone or the paved gutters, while the rest of the surface is compacted by the pilon, beginning with light blows and ending with vigorous ones. All these tools are heated *nearly* red hot, as the powder sticks to them when they are cold. After the surface has been thoroughly compacted it is tested with a straight-edge, and then rubbed with the lissoir, also heated, giving a glaze to the surface, after which it is dusted over with cement and allowed to cool thoroughly before the traffic is turned on it.

Efforts have been made to secure a more regular surface than is practicable by pilonnage and at less expense by rolling, but it was difficult to so heat the rollers that the powder would not adhere, and the pavement was liable to tear as it was about finished.

The powder retains heat for some time, and the work goes on in a continuous sheet for the day, a gang of ten or twelve men being able to complete about 600 square yards per day. In the morning the uncompressed powder at the edge of the work of the day before is swept away, and hot powder put in its place, which in turn is removed after having heated the old work ; the spreading and pilonnage then goes on as before, leaving small or no trace of the junction.

Great care should be taken that the concrete foundation is thoroughly dry, otherwise the hot powder evolves steam, which permeates the powder

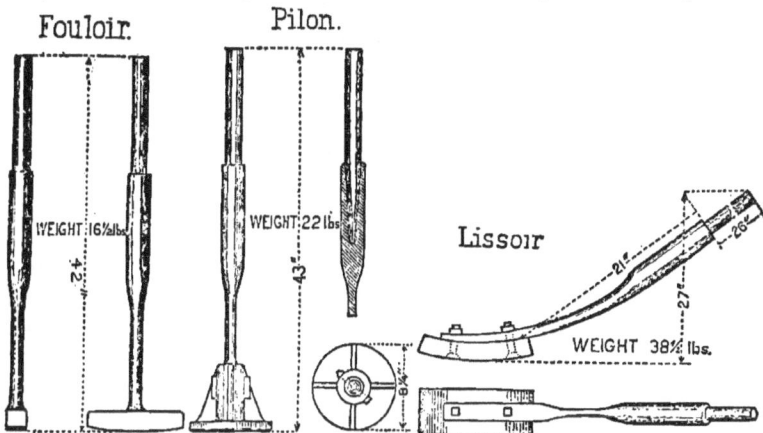

Fouloir.   Pilon.

WEIGHT 16½ lbs.   WEIGHT 22 lbs.

Lissoir

WEIGHT 38½ lbs.

and leaves the compressed mass in nodules called *almonds*. These do not show during pilonnage, but are developed by the traffic, when the place is cut out and refilled. A very slight defect seems capable of starting a hole in compressed asphalte, and for the first two weeks it is under traffic it should be watched closely and repaired promptly. The defects caused by deficient aggregation of the molecules show themselves more readily than those caused by steam. Overheating, which renders the asphalte as inert as sand, is one cause of deficient aggregation, and laying the powder too cold is another.

When the asphalte is too rich in bitumen or the bitumen is too oily, the compressed asphalte forms waves under the traffic, sometimes longitudinal and at others transverse.

So far but few asphaltes have been found that are available for compressed pavements. The Val de Travers (from Neuchatel) was first used,

but in 1867 Seyssel (Pyrimont) was used with success in the Rue de
Richelieu, and both have since been used by the General Asphalte Company of Paris, which, until January, 1878, laid and maintained the asphalte
roads in that city. The Val de Travers asphalte is of the two more regular in its grain and impregnation, and richer in bitumen, having 9 to 13
per cent. Seyssel is not so regular in its grain and impregnation, and
contains from 7½ to 10 per cent. of bitumen. The base of both is equally
pure carbonate of lime, containing about 2 per cent. as a maximum of
silica. Of the two, M. Ernest Chabrier, for a long time manager of the
Paris Company, in his paper read before the Institute of Civil Engineers,
says: "No engineer could conscientiously say that the Val de Travers is
better than the Seyssel asphalte. The former may be safer in the execution of a work not subjected to supervision ; the latter offers greater guarantees of good execution, because more care is required in the work."

It was also held that, for compression at least, the two should not be
mixed, and that only pure limestone, impregnated with from 8 to 12 per
cent. of bitumen, was available for compression ; but within the last two
or three years the Limmer & Vorwohle Asphalte Company of London
found their mastic pavement breaking up and employed the Sicilian
asphalte (which is certified by W. J. Fewtrell, F. R. S., to contain as
high as 30 per cent. of bitumen) in compressed pavements and sidewalks.
Aldgate and Newgate streets having been laid with this asphalte are
apparently wearing well. And in Paris, Paul Crochet, who has the
contract for new work and maintenance for five years from 1878 (see
Appendix, No. 4) has used, so far, a mixture of Lobsan and Seyssel
Forens-Nord. M. Malo says of Lobsan : " It contains within its bitumen
an oily substance which renders it too soft and injurious to the consistence of asphaltic mastic ; we free it from this oil by distillation, after
which it is in a proper state to be used." The Seyssel Forens is a heavy
limestone, very poor in bitumen. As after a careful investigation the
engineers of the Ponts et Chaussées have allowed it to be used in Paris
it seems that, theoretically, at least, a mixture of two dissimilar asphaltes
is not disadvantageous, nor is the use of an asphalte that contains a hydro-
carbon of the more volatile series ; also that, as in the case of the Sicilian,
an asphalte containing more than twice as much bitumen as was formerly
thought the safe maximum can be successfully laid under exceptionally
heavy traffic.

There was an unwillingness on the part of the manager of the Limmer
& Vorwohle Company to explain the manner in which he handled his
material, but it is probable that it was skillfully roasted so as to drive off
the excess of bitumen as his compressed roadways showed less tendency
to roll in hot weather than those laid from Val de Travers rock.

By the kindness of Count Kielmansegge, one of the directors of the

Neuchatel Company, I am able to give the following analyses of Val de Travers asphalte. These two sets of analyses by Charles Heisch, F. C. S., were made in consequence of complaints by their customers that the rock was becoming too rich in bitumen:

	September, 1876. 5 samples.				
Bitumen	9.46	9.48	11.98	11.96	10.06
Carbonate of lime	90.54	90.52	88.02	88.04	89.94
	100.00	100.00	100.00	100.00	100.00
Loss at 212°	1.46	1.42	3.04	3.22	1.92

	July, 1877. 3 samples.		
Bitumen	10.2	11.5	12.32
Carbonate of lime	89.8	88.5	87.68
	100.0	100.0	100.00
Loss at 212°	2.46	2.6	3.2

The loss was moisture and light oils. A small amount of silica was estimated with the lime.

Mr. J. Knight, Secretary Society Francais des Asphaltes, gave me the following analyses made by Steen, of Copenhagen:

Seyssel Pyrimont	Organic...... 12.00	Soluble in ether......	9.62
		Insoluble in ether....	2.35
	Inorganic..... 88.00	Soluble in acid.......	86.45
		Insoluble in acid.....	1.55
Seyssel Garde bois..... ..	Organic 9.00	Soluble in ether......	9.15
		Insoluble in ether....	.55
	Inorganic. ... 90.30	Soluble in acid.......	85.10
		Insoluble in acid.....	5.20

The Garde bois asphalte is only used as a mastic, according to Mr. Knight, not having coherence enough for satisfactory compression.

Mr. W. H. Delano, manager of the Compagnie Generale des Asphaltes de France, says of an asphalte for compression that " it should be carbonate of lime (without admixture of foreign material) and mineral bitumen which does not evaporate at 302 degs. F. Some bitumens contain an oil which commences to evaporate at 158 degs. F.; they may, when purged of the light oil, leave an excellent quality behind. The physical qualities of the limestone should also be carefully examined ; but whatever the result of even a careful analysis may be, the value of a new asphalte cannot safely be determined except by actual experience extending over three years, at least, of hard winters and hot summers." Col. Haywood also was understood to say that at least three years were necessary to prove the good qualities of a new or untried asphalte. His experience is strongly against any form of asphalte pavement except " compressed."

From a sanitary point of view an asphalte pavement is without a peer ; its surface is smooth, regular and non-absorbent, with no cavities or cracks of any kind to retain the infected mud and dust of the streets, and the soil beneath it is kept dry. It is more thoroughly cleaned, either by sweeping or washing, than any other pavement. Its freedom from noise and other excellencies is fast placing it in all the business and banking streets of the City of London, where it seems to be superseding all other pavements.

In comparison with granite pavements, its great economy is to brain workers and the owners of horses. M. Darcy estimates that the expense of maintaining and renewing horses and carriages in Paris would be reduced one-half if asphalte was substituted for pavement throughout the city.

Its disadvantages are, it is expensive ; it is not well to lay it on a grade much greater than two per cent.; it is subject to slight decay in the gutters, which they try to correct in Paris by laying the gutters with stone sets ; and lastly, the most serious charge, that it is slippery and dangerous to horses, No doubt at the commencement of a rain, or in foggy weather, its surface is slippery unless it is very clean, and a horse turned suddenly is liable to fall; but when dry or thoroughly wet no such charge can be brought against it.

Referring to Col. Haywood's much quoted tables comparing granite, asphalte, and wood, we find that before an accident occurs a London horse will travel on granite 132, on asphalte 191, and on wood 446 miles. "That of those accidents which are most obstructive to the traffic, as well as most injurious to the horses, asphalte had the greater proportion, granite the next and wood the least."

He says further in a report upon asphalte and wood pavements :

First.—As regards convenience : That asphalte is the smoothest, driest, cleanest, most pleasing to the eye and the most agreeable pavement for general purposes, but wood is the most quiet.

Second.—As regards cleansing : That wood may be kept cleaner than it hitherto has been, but will be more difficult to cleanse effectually than asphalte. That as both pavements require occasionally strewing with sand or gravel, there is no difference between them in that respect.

Third.—As regards construction and repairs : That asphalte and wood, taking all seasons into account, can be laid and repaired with about equal facility ; but the smallest, neatest, cleanest and most durable repairs can be made in asphalte.

Fourth.—As regards safety : That, whether considered in reference to the distance that a horse may travel before it meets with an accident, or the nature of the accident, or the facility with which a horse can recover its footing, or the speed at which it is safe to travel, or the gradient at which the material can be laid, wood is superior to asphalte.

Fifth.—As regards durability and cost : That wood pavements. with repairs, have, in this city, had a life varying from six to nineteen years, and that, with repairs, an average life of about ten years may be obtained ; that the durability of the asphaltes is not known, but that under the system of maintenance adopted they may last as long as wood ; that, contrasting the tenders for laying and maintaining for a term of years the two best pavements of their kinds, wood will be the dearest.

The above remarks refer to London, and inasmuch as our climate is much the driest, and as asphalte is safer where dry, the comparison would be more in favor of asphalte here than there.

Both the compressed and mastic surfaces are easily and thoroughly cleaned by washing and by sweeping when either wet or dry. A *squeegee*, which is a long-handled scraper with a rubber blade 32 inches long, when pushed along a wet pavement both cleans and dries it.

It may at first sight seem that if a person will indulge in the luxury of changing his gas company on the recipt of each bill he should be compelled to make good the surface of the street, and the ease or difficulty of making repairs should largely be a consideration for those who open the streets, but as a matter of fact it is almost impossible to properly inspect the work of plumbers, and while one avenue in this city has on its surface a double-tracked horse railroad and 240 elevated railroad columns per mile, with one sewer, three water pipes, and eight gas pipes (belonging to four different companies) under it, the subject cannot be without interest. In general, any pavement with a foundation of concrete can be, and probably will be, more thoroughly repaired than one with a loose foundation, such as sand, broken stones, or Telford. For the surface, asphalte is superior to all others in the matter of neatness, Macadam next, and tne granite and wood pavements next.

ASPHALTIC MASTIC.

The asphaltic mastic, which is greatly used in Paris for sidewalks, is made, as above defined, from asphalte and bitumen. This bitumen was at first obtained from the asphalte and molasses of Seyssel by boiling either the broken stone or sand (of the molasses) when the bitumen separated, and was skimmed off the surface of the water and sides of the boiler. This bitumen (after being re-melted and allowed to settle to clear it from sand, if necessary) is melted in an iron boiler to the amount of between 5 and 10 per cent. of the required batch, and the powdered asphalte added slowly and stirred in until all is thoroughly incorporated. It is then cast in molds for sale or use as mastic. When this mastic is used it is necessary to break it up and add it again to melted bitumen, after which 50 to 60 per cent. of fine or coarse silicious sand is added, and the whole kept heated till it does not stick to a clean wooden stirrer.

The mastic is then taken to the place where it is to be laid, in a *locomobile*, essentially a horizontal cylinder on four wheels, with a fire-box under it and an arrangement for stirring the mastic so as thoroughly to mix the gravel and to keep the mixture from burning. It is then laid on a concrete foundation, generally for sidewalks ½ inch thick, or more. It is poured on the concrete from conical sheet-iron pails, and spread by the aid of a wooden hand float. After the spreading is completed, sand, and in some instances gravel the size of peas, is added, and the surface well rubbed. The sand adds to the wear of the sidewalk, and prevents the mastic softening under the heat of the sun. The mastic is sometimes laid in one course and sometimes in two.

M. Malo insists on the bitumen used for mastic being the same as that impregnating the limestone, but, on account of its scarcity, it has been almost entirely replaced by refined and tempered Trinidad or other bitumen.

Crude Trinidad contains, besides water, chips, leaves, etc., twenty-five to thirty per cent. of clay, in which state it is brittle, and the clay, as in the case of compressed asphalte, is an element of weakness, as it attracts water. It is refined, according to Malo, " with goudron," a product of a second distillation of schists, which, at a temperature of 59 degrees to 68 degrees Fahrenheit, has the aspect of a completely liquid bitumen ; 600 to 700 pounds of this goudron is boiled in a chaldron, into which the crude bitumen, broken to the size of an egg, is thrown in successive doses until the weight is 1,700 to 1,900 pounds. After the ebullition has ceased the fire is drawn and time allowed for the sand and earth to settle, when the purified bitumen is drawn off and strained. The mineral tar of Autun is said to be the best substance now in the market for refining and tempering bitumen.

M. Malo deprecates the use of oils from boghead coal and petroleums generally. All of the manufacturers interviewed on the subject denied the use of still bottoms or petroleums, and most agreed with Malo that boghead oils should not be used, on account of the cracks which occurred in the mastic ; others claimed that the cracks were due either to a poorer quality of asphalte than they used, or to an insufficient purification of the Trinidad. See also App. No. 4, art. 17, and App. No. 5, art. 21.

There is a great difference between the true asphaltic mastics, depending on the purity and quality of the asphalte, the purity and component parts of the goudron, and the skill of manipulation. A well-made mastic is decidedly harder than compressed asphalte, which to a certain extent is partially fluid under pressure, and increases in density under traffic much faster than the surface is abraded. Mastic, on the other hand, as above stated, is hard, crumbling under too great pressure, not noticeably compressible, but disintegrating under the blows of horses' feet. It, however, has much more cohesion than compressed asphalte, and should be preferred in those situations where it is not liable to receive the compressive action of heavy loads, and is exposed in large areas to changes of temperature ; the partial fluidity of asphalte keeps such cracks closed under traffic.

In the courts of the new House of Parliament, eighty by forty-six feet, Claridge's mastic stands without cracks, a beautiful surface ; while carefully laid compressed asphalte in the court next to it has cracks in its surface. Claridge's asphaltic mastic, which has a very high reputation in London, is made from Seyssel asphalte and a very carefully refined Trinidad goudron. It is claimed that the Trinidad bitumen, when thoroughly

cleared of its clay, has the greatest cohesive strength of any of the bitumens, and the Claridge Company claim that they conduct their refining
processes with more skill and thoroughness than any other company.
Their mastic is laid in two courses in strips about 3 feet wide, breaking
joints.

It is believed that in Paris no compressed asphalte is laid for sidewalks, but in London the Limmer Company is compressing Sicilian on
sidewalks, and the Val de Travers Company always use compressed.

Some mastic was laid in London streets, but it is now being rapidly
repaired or renewed with compressed asphalte, except in narrow streets
of small traffic, nor is asphaltic mastic laid in the wheelways of Paris.

There is one theoretical point in regard to different asphaltes to which
the attention of the Society may be called. That is the fact that some
asphaltes will compress and others will only make mastics. The Val de
Travers and the Seyssel are very good for either purpose. Lobsan is
capable of being compressed, and with proper treatment, making a mastic.
Sicilian, as far as known, has only been employed for compression. The
asphaltes of Hanover and Brunswick, while they make very good mastics,
seem incapable, with the present knowledge, of being compressed, and
the writer has met no one who could explain the cause of the differences.
It evidently does not lie in the quantity of bitumen contained in the rock,
as some Hanover asphaltes contain more than either the Seyssel or the
Val de Travers.

BITUMINOUS MASTIC.

Bituminous mastics, or the fictitious mastics of Malo and the French
writers, were experimented with very generally in Europe, in the hope of
finding something as useful as asphalte at a less price ; but it was found
that no practicable degree of heat and pressure would give the microscopic
impregnation of grain found in the natural mineral.

In this country the effort took a different direction. Refined and properly tempered Trinidad bitumen was mixed with limestone or fine or coarse
sand, and a product was sought between asphalte and asphaltic mastic.
It is applied neither as a dry powder nor in a fluid or semi-fluid condition,
but as a sticky and coherent mass, and is subjected to pilonnage and rolling. While it does not seem to have either the ductility under traffic or
the resistance to wear of asphalte, it has less hard brittleness than the
true asphaltic mastic ; its great difficulty being a tendency to disintegrate
in a rotten manner, possibly from the use of improper oil in refining and
tempering the Trinidad bitumen, or else from the inherent difficulty of
making an artificial mixture equal in quality to the natural product.

The most successful pavement of this class in this city is the block on
Fifth avenue opposite the Worth Monument, laid by the "Grahamite
Company." Its component parts are understood to have been a mixture

of Trinidad bitumen with fine sea sand containing a little carbonate of lime, and enough Grahamite or Ritchie mineral " to save the patent." This pavement was laid in the spring of 1873, and after the rectification of surface generally necessary, it stood till the fall of '77 without requiring any repairs that could be charged to its own weakness. It was repaired by another company in the fall of 1878, and now requires additional attention. During this time it has filled all the conditions demanded of a good pavement to a greater extent than any other pavement in the city. Three other pieces laid by this company have not proved so successful.

Pavements of this class are very fully described in General Gilmore's Roads, Streets and Pavements.

They are being extensively laid in Washington under a contract for three years' maintenance. As in France and England, it is held that whatever the physical properties and composition of an untried asphalte may be, it is not safe to pronounce it a success until after three years' wear; it seems that a longer maintenance contract would be advisable.

In London, Colonel Haywood, as noticed under the head of Wood Pavements, lets his asphalte pavements at a certain price per square yard for laying, and a provision that they shall be maintained for 2 years at the contractor's expense and 15 years at an agreed price per square yard per annum, and a good pavement be given to the city at the end of the 17 years.

Several different kinds of pavements, compressed and fictitious and other mastics, were laid, the Val de Travers compressed asphalte being the only one that was satisfactory, and most of the other pavements were relaid by that company at the expense of the sureties. The Limmer & Vorwohle Company, however, introduced the use of Sicilian, with which they have relaid Newgate and Aldgate streets, and are relaying Cornhill. ·

In the City of Jassy, Roumania, the contract (see Appendix No. 5) is based on the London plan. The work was commenced in 1873 and finished in 1878. It is to be maintained at the contractor's expense until 1880, from that time, and until 1895, at a stipulated price.

In Paris (see Appendix No. 4) the system is to let the construction, maintenance, and repairs of pavements and footpaths for a term of years, 5 to 10, without guarantee of wear, relying on rigorous specifications and thorough inspection for the excellence of the work.

The plan pursued in London and Jassy seems preferable, even with the natural asphaltes, for after the solidity of the foundation is assured, the duration of the wheelway depends so much on the technical skill and thorough honesty exhibited in small details, that the burden of proper inspection should be thrown on the contractor by his having a large pecuniary interest in the soundness of the work done and material furnished. The inspection, however, should be close enough to keep the engineer in charge fully informed of the methods and materials employed, particularly towards the close of the contract. But no greater injury could be done to a city's system of streets than strictly enforcing the specifications of some of the 300 and more patents taken out for mastic pavements.

The possibility of having any street in continuous good order for 17 years is probably too much for the imagination of the average New Yorker. And no one can well realize the effects of having all the streets so kept in repair.

TAR MASTIC.

Among the legacies of the " ring" were the various tar pavements. They were worse than the wood pavements, though not being so thick, they have not proved so enduring a nuisance as the latter.

The best of them are understood to have been made of crude coal tar mixed with gravel and sand, and on the evaporation of the light oils they went to pieces, generally with great celerity. Occasionally a patch would be found that stood the traffic pretty well; this was supposed to result from the evaporation of the oils before mixing with the gravel. G. Leverich, Member of the Society, proposed to secure a uniform product by passing steam of a certain temperature through the tar and driving off all the oils of low specific gravity. Coal tar is now generally reduced to pitch to get the aniline colors, their most valuable product, and the pitch may be tempered by melting with dead oils, a waste product of the distillation. It, however, is very brittle, and it is doubtful if it can be employed economically, except for side-walks of light traffic.

APPENDIX NO. 1.

Extract from Specification of W. S. Till, Borough Surveyor, Birmingham, 1877.

The whole of the carriage way to be excavated to the required depth, and the foundations formed to uniform gradients and proper cross sections, with ashes or other approved materials. When these have been rolled and are well set, the contractor to coat the whole width and length of the carriage way with not less than 8 inches of good, strong, clean gravel, screened through a ¼-inch riddle, or with not less than 8 inches of approved slag, broken regularly to such a size that the largest piece, on its greatest dimension, will pass through a 3-inch ring.

The gravel or slag to be well gritted and watered, and kept raked together and rolled until the whole is consolidated.　　*　　*　　*　　*　　*　　*　　*　　*

The contractor to coat the carriage way, throughout its whole length and width, with not less than 6 inches of Rowley Ragstone, broken to pass through a 2½-inch ring.

The stone to be gritted, watered, and kept together, raked and rolled until the whole is consolidated. Crown 6 inches above curbstone. Each layer to be laid on complete throughout the whole length and width of carriage way, and each layer to be separately rolled with a heavy roller.

APPENDIX NO. 2.

Extract from Specification for Telford, Macadam and Trap Block Gutters on Fifth Avenue.

(4). DESCRIPTION OF MATERIALS.—The stone blocks are to be of double and uniform quality, each measuring, on the face or upper surface, not less than 4 nor more than 8 inches in length, and not less than 4 nor more than 6 inches in width, and in depth not less than 6 nor more than 8 inches; blocks of 4 inches in width on their face to be not less than 3 inches in width at the base; all other blocks on transverse measurement on the base, to be not more than 2 inches less than on the face, but no block on the base shall be of less width or length than 3 inches, and to be in all respects equal to the specimen blocks at the office of the Commissioner of Public Works.

It is also required that the sides of the stones (which form the joints) shall be so sufficiently even and properly shaped that joints may be formed with a similar side not exceeding ¼ of an inch, no stone having an objectionable protruding face will be accepted or allowed to be used for paving. They will be carefully inspected after they are brought on the line of the work, and all blocks which, in quality and dimensions, do not conform strictly to these specifications, will be rejected, and must be immediately removed from the line of the work. The contractor will be required to furnish such laborers as may be necessary to aid the inspector in the examination and culling of the blocks; and in case the contractor shall neglect or refuse so to do, such laborers as in the opinion of the Commissioner of Public Works may be necessary, will be employed by the said Commissioner, and the expense thus incurred by him will be deducted and paid out of any money then due or which may hereafter grow due to the said contractor under this agreement. The blocks must be of trap rock or of syenite, equal in hardness to what is called Quincy granite.

TELFORD.—The stone for the foundation of the pavement is to be sound, hard and durable quarry stone, each from eight to ten inches in depth, from three to six inches in width, and from six to fourteen inches in length, and of a sufficiently uniform size to be acceptable; except the stone under the gutters, which only differ from those above described in that they must be not less than six inches in depth, and except the course of stone under the Macadam, which is next to the gutters or the bridge stone, which stone is to be twelve inches in depth, and in every other respect as first above described.

The broken stone for the bottom course is to be of trap rock, and of such size that would pass in any direction through a ring with interior diameter of $2\frac{1}{4}$ inches, and of a sufficiently uniform size and proper shape to be acceptable, to be composed only of stone that is hard and durable, and sufficiently free from screenings, and practically free from dirt and other foreign matter. The broken stone composing the next overlying course is to be the same in every respect as just described, except that each stone is to be of such size that would pass in any direction through a ring with interior diameter of 2 inches.

The coarse screenings are to be those of trap rock ; the stone is to be practically free from dirt and other foreign matter, to be composed of material only that is hard and durable, no particle of which is to be of larger size in any direction than $1\frac{1}{4}$ inches, and all the particles to be of such relative size to $1\frac{1}{4}$ inches as shall be acceptable; through these screenings is to be evenly mixed before rolling a sufficient quantity of the finer screenings of trap rock to form a proper and secure bond.

The fine screenings of trap rock are to be practically free from dirt and other foreign matter, and to be composed of material both as to size and quality that will be acceptable.

(9). PREPARATION OF ROAD BED, &c.—All paving and other stones necessary to be removed shall be taken up and immediately removed from the line of the work ; the Belgian and granite blocks to be deposited where directed by the Water Purveyor for the use of the city; the subsoil or other matter (be it earth, rock or other material) shall then be excavated and removed, to such depth to be determined by the engineer in charge, that when rolled and finished, irrespective of the covering of fine screenings, the pavement throughout its entire extent shall be at least sixteen inches thick. Should there be any spongy material or vegetable matter in the bed thus prepared, all such material shall be removed, and the space filled with clean gravel or sand carefully rammed or rolled, so as to make such filling compact and solid. No ploughing will be allowed in preparing the foundation.

The road bed shall be truly shaped and trimmed to the required grade, and with such crown as shall be directed by the said engineer, and rolled with a roller weighing not less than two tons, until the surface is firm and compact.

LAYING THE FOUNDATION OR PAVEMENT.—After the road bed is prepared, agreeably to the terms of this specification and to the satisfaction of the engineer, and the stone hauled and deposited thereon, the foundation stones shall be laid by hand in the form of a close, firm pavement. They shall be set on their broadest edges and lengthwise across the road, except in the case of the stone under the Macadam, which is next the gutter stones or bridge stones, which foundation stone is to be placed with its longest side parallel to the curb or bridge stone; after being set closely together, they are to be firmly wedged by inserting and driving down with a bar, in all possible places between them, stones, as near as practicable, of the same depth, until all the stones are bound and clamped in proper position; all the projections and irregularities of the upper part of the pavement shall then be broken off with a hammer, care being taken not to loosen the pavement, and the spalls and chips are to be worked and driven with the hammer into all the interstices not already filled by the process of wedging, so that the pavement, when completed, shall present a sufficiently even surface, and be at each point of such thickness as required by these specifications. No wedging shall be done within twenty-five feet of the face of the work that is being laid, and the stone foundation must be in a compact and satisfactory condition in every respect at the time of the spreading of the broken stone.

MACADAMIZING.—After the stone foundation has been completed agreeably to these specifications, and has passed the inspection of the said engineer, a layer of broken stone of the quality and size herein specified for the bottom course, and of such a depth as will make 4 inches when rolled, shall be spread evenly over the pavement ; this layer is then to be rolled until sufficiently compact.

The next overlying course will be of stone, as hereinbefore described for said course, and is to be spread so that the surface will be uniformly 1 inch below the grade and crown when the pavement is finished.

Stone hammers are to be used on this course, so that when rolled as much as shall be required the surface of each stone that is exposed will not have a longer dimension in any direction than 1 inch.

A layer consisting of the coarse screenings herein specified is then to be applied and spread to such depth as will bring the surface to the proper grade, irrespective of the finish of fine screenings; this layer is then to be rolled, and during the progress of the rolling, if necessary, coarse screenings shall from time to time be applied, so that when the rolling ceases the roadway is truly surfaced to the required grade and crown. This layer is to be rolled until all settlement ceases and the stones are thoroughly compact and the surface true to the grade and crown.

During the process of rolling any course of stone there shall, when required, be spread lightly over the same from a shovel, the fine screenings of trap rock herein described; and after the upper layer has become thoroughly compact, there shall be spread upon the surface fine screenings, so much as to produce a covering half an inch in depth when rolled, and the rolling is to continue until, by a sufficient use of water, a wave is produced before the wheel of the roller.

The rolling of the Macadam stone and screenings shall be done with a roller weighing not less than any of the steam rollers in the possession of the Department of Public Works.

Each layer of broken stone and the screenings or binding material shall be well and thoroughly rolled, and the rolling upon each layer shall be prosecuted until, in the opinion of the engineer, each course shall have been completed, as hereinbefore specified, and until each layer and the finished surface shall be rolled and finished to his entire satisfaction and approval.

While the rolling of each layer of broken stone and the screenings or binding material is being presecuted the work shall be kept moistened to such extent and in such manner as required, and care shall be taken that too much water is not applied while rolling the first layer, and until after the interstices are well filled with the binding material.

The pavement, when completed, shall be at each point of such construction and at least of such a depth as required by the specifications, and of such crown and such form of gutter as shall be directed, and in any case the thickness of the pavement is to be determined on a line at right angles to the grade and crown.

The use of a proper roller, rammers, or other suitable implement, is to be substituted for that of the steam-roller when necessary.

The construction of the foundation stones, and the Macadam pavement shall proceed so as to be practically equally as far advanced across the entire width of said pavements.

Particular care and attention will be required in obtaining a satisfactory joining of the Macadam paving and the blocks in the gutters.

PAVING THE GUTTERS.—Upon the stone foundation already described, shall be laid a bed of clean, sharp sand, not too fine, or clean fine gravel, of the depth necessary (about 4 inches) to bring the paving to the proper shape and grade when rammed. The stone blocks are to be laid with joints at right angles to the curb, with joints not exceeding three-quarters of an inch, at such grade and in such form as shall be directed; each course of blocks shall be of a uniform width and depth, and so laid that all longitudinal joints shall be broken by a lap of at least one inch; and the blocks next the curb shall break joints with the curb by at least one inch; as the blocks are laid the joints shall be so filled with sand or gravel as to secure, when the work herein mentioned is completed, against there being any spaces not filled with sand or gravel between the blocks; they shall be covered with clean, sharp sand, which shall be raked until the joints become filled therewith, the blocks shall then be thoroughly rammed to a firm unyielding bed, with a uniform surface to conform to the grade and crown of the avenue, as shall be directed. No ramming shall be done within 25 feet of the face of the work that is being laid; and the ramming is to be done at such times and in such manner with reference to the rolling as shall be directed. When the paving from the centre of an intersecting street to the centre of the next intersecting street is constructed, it shall be covered with a good, sufficient second coat of clean sand, and shall immediately thereafter be thoroughly rammed until the work is made solid and secure, and so on, until the whole of the work embraced in this contract shall have been well and faithfully completed in accordance therewith.

This portion of the work, laid with trap blocks for the gutters, will be measured and included in the returns for Macadam pavement.

No stone or other material, except sand or granite, is to be placed on the blocks that are paved, until these are rammed. When necessary, in order to make good joints, the blocks are to be trimmed down on the sides by and at the expense of the contractor.

APPENDIX NO. 3.

ROAD ROLLERS.

As far as known to the writer, the 4 rollers here described are the only ones in the market ; they all run with equal facility with either end forward.

THE GELLERAT ROLLER.

These machines consist essentially of a locomotive boiler, supported on a frame which is carried on two cast-iron rollers, each 3 feet 11 inches in diameter, and 4 feet 7 inches long for the 15-ton rollers, and 4 feet 9 inches in diameter and 6 feet long for the 30-ton rollers. Their peculiarity

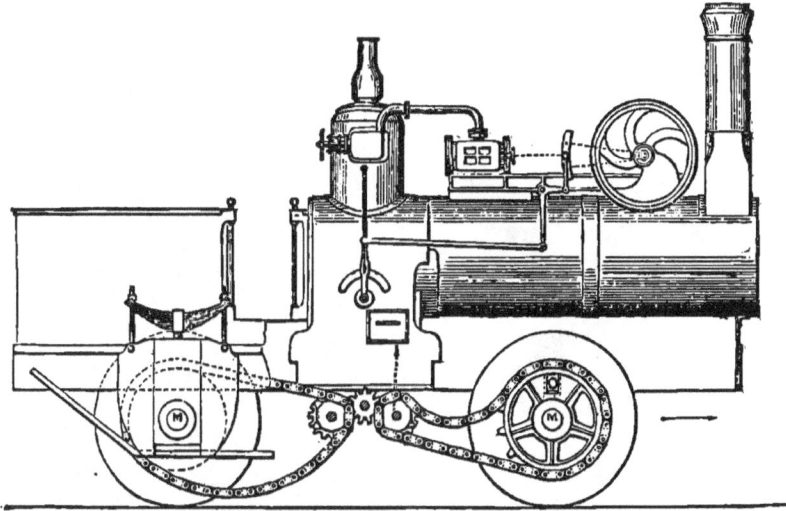

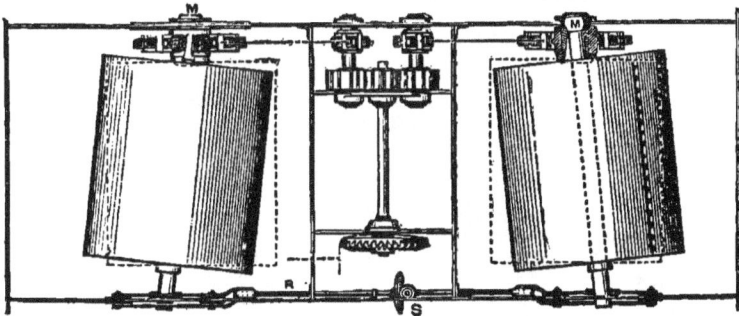

THE GELLERAT ROLLER.

is that each roller is a driving wheel, and bears half the weight of the machine, which is guided by a hand wheel working from the foot-board into a bevel gear *S,* Fig. 2, which works the right and left hand screws, *R R,* throwing the axes of the rollers into radial positions; the other, or driving end of each axle, is stationary, with a spherical bearing—Figs. 1 and 2.

THE LINDELOF ROLLER.

This roller has an upright boiler, and two vertical cylinders that actuate a beveled gear which works into a gear bolted on to the driving wheel, the tread of which is a 1-inch plate of wrought iron ; the driving roller bears two-thirds of the weight of the machine. Both 10 and 15-ton rollers

THE LINDELOF STEAM ROAD ROLLER.

are built ; for the 10 ton, of 2,000 pounds each, the driving roller is 6 feet and the steering roller 5· feet long, giving a weight on the driving roller per inch run of 185 pounds.

THE AVELING AND PORTER ROLLER.

This roller, which is extensively used in this country, has been improved in several particulars, and is believed to be the only roller now in the market that is run and fired by one man.

As now manufactured the outside wheels are the driving wheels, the steering wheels covering the space between them. The boiler is horizontal and multitubular ; the single steam jacketed cylinder is on top of the boiler, and runs a fly-wheel which by the aid of gearing drives the roller at a speed of about 2 miles per hour. The driving wheels have holes in their treads in which spikes may be placed for tearing up the road bed before remetalling, and the fly-wheel makes the roller available as a stationary engine to run a stone breaker.

Four sizes are manufactured, viz., 8, 10, 15 and 20 tons in weight. Two-thirds of the weight is carried on the driving wheels, which have a width of 2 feet 2 inches for the 20-ton roller, giving

THE AVELING & PORTER ROLLER.

a weight per inch run of 574 pounds ; the width of driving wheels on the 15-ton roller, is 1 foot 10 inches, giving a weight per inch of 509 pounds. The roller is fitted with a friction brake.

THE ROSS ROLLER.

This machine is a combined road roller and rammer. Only one size is built, weighing 44,000 pounds. The boiler is vertical and the rams, any or all of which can be used or not, are actuated

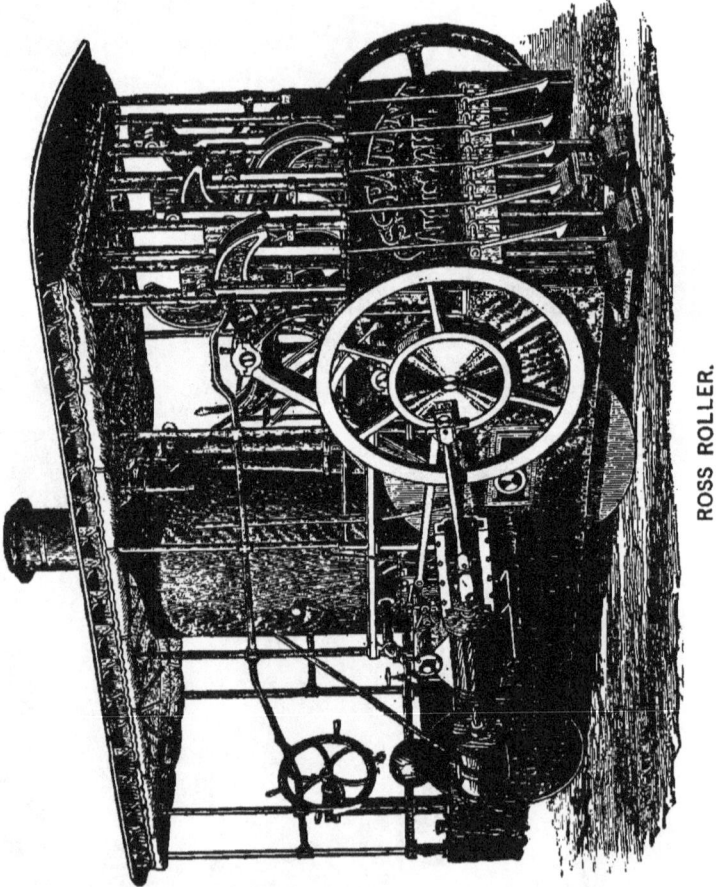

ROSS ROLLER.

by cams. The length of the driving roller is 6 feet in all, with a space in the middle of 8 inches for the driving chain. The steering roller is 80 inches long. Four-fifths of the entire weight is carried by the driving wheel, giving a compressive force per inch run of 550 pounds. It is claimed that the traction is sufficient to allow it to mount grades of 20 feet per 100. Maximum speed, 5 miles per hour.

The rams, five in number, are said to give an effective blow of 7,000 pounds each ; they are of no use in compressing and puddling trap, but are efficient with limestone, and must be efficacious for rubble foundations. The frame is utilized as a water tank, and the driving roller can be heated by steam for rolling mastics, &c.

APPENDIX No. 4.

A part of the Contract for the Construction and Maintenance of the Streets and Sidewalks of the City of Jassy.

Between M. Nicholas Ganć, Mayor of Jassy, and Mr. W. O. Callender, of London.

ARTICLE 1. The contract has for its objects :

(*a.*) The construction of a system of streets with gutters and a system of sidewalks with curbs.

(*b.*) The maintenance of the streets and sidewalks constructed under this contract.

ART. 2. The construction consists of the following items of work :

(*a.*) The construction of 47,853.8 square yards of streets, covered with compressed asphalte.

(*b.*) The construction of 179,384.3 square yards of sidewalks, covered with asphaltic mastic.

(*g.*) The construction of 17,039.4 square yards of roadway, paved with Macadam made from stone taken from streets now paved.

(*h.*) The construction of 49,210 lin. feet of granite curbing for bordering the asphaltic mastic sidewalks on streets covered with compressed asphalte.

ART. 3. The time fixed for the completion of this contract is five years, to begin with July 1st, 1873, and to end on March 31st, 1878.

ART. 5. The labor of maintenance will consist in maintaining the streets in a constant good condition. On all the streets and sidewalks constructed by him, the contractor will repair, with his own laborers and his own material, all degradations as soon as they appear, and will replace the material lost.

ART. 6. The length of the contract for maintenance is fixed at 15 years, to begin on May 1st, 1880, and to end on April 30th, 1895.

ART. 10. The transverse inclination of the asphalte sidewalks will be 2 per cent.

ART. 11. The curbs of sidewalks of asphalte, on asphalted streets, will be of granite; those of sidewalks of asphalte, on streets paved with blocks or cobble stones, will be of sandstone, and the curbs of Macadam steeets, when they protect sidewalks of the same material, will be of stone cut from the old flagstones of the present sidewalks.

The curbstones will be of the following dimensions : The faces parallel to the surface of the sidewalks will be 5.5 inches in width, of which 0.9 inch of the upper face will be eventually covered with asphalte. They will have a height of 11.9 inches, of which 6 inches will be under ground and the remainder above the gutter ; their length will not be less than 15¼ inches. The form of the curb will be that of a parallelopipedon, having next the sidewalk a rebate of 0.9 inch in width and a height equal to the thickness of the bed of asphalte.

ART. 14. The gutters of asphalted streets shall have, according to the locality, a maximum width of 3.3 feet ; they shall be constructed of new cobble stones, well culled, or of new sandstone blocks. At the edge of the asphalte there will always be laid a range of granite blocks 10.6 inches in length, 5.3 inches in width and 9 inches in depth.

ART. 15. The sandstone blocks for the gutters will be dressed to regular cubes in form, of 9 inches on each side.

ART. 16. The asphaltic rock, which must be natural and not artificial, must be from the best asphaltic quarries known: Limmer, in Hanover, Seyssel, in France, and Val-de-Travers, in Switzerland. The asphaltic mastic to be used must satisfy the following conditions :

(*a.*) It must contain 12 parts of bitumen to 88 parts of asphaltic rock.

(*b.*) The asphaltic rock itself must contain at least 7¼ per cent. of bitumen, and at the most 93 per cent. of pure carbonate of lime.

(*c.*) Rocks which contain, even in small proportions, quartz, sulphates, iron pyrites or aluminum must be rejected for the composition of mastic.

ART. 17. The bitumen used for mastic must be natural, not artificial, from the bitumen lakes of the Island of Trinidad, or from the asphaltic rocks of Seyssel, if it is proved to be equal in quality to the Trinidad bitumen : it must be free from water ; its specific gravity must be from 1.1 to 1.5. Dipped in water at the freezing point, it must not lose its ductility ; its surface must present no cracks or streaks. The surface of a fracture must be black and brilliant. It must be perfectly soluble in petroleum oil or the spirits of turpentine, and the solution when passed through a filter must leave no residue. (?)

ART. 18. The quarry whence comes the bituminous rock, the mastic and the bitumen, must be certified by authentic certificates to be of even fabric and its products of the first quality. Each block of mastic must bear the trade mark of the quarry.

ART. 19. The contractor is forbidden to have in his storehouses any asphaltic rock or mastic which do not comply with the conditions above stated and not in conformity with the samples deposited with the Mayor on the day of the signing of this contract; he must neither use nor have in his storehouses any bituminous or resinous oils other than those specified in Article 17.

All other material, resinous, bituminous or oleaginous and all other mastic than that like the samples deposited with the Mayor that may be found in the storehouses of the contractor will, on the first offense be confiscated, and a fine of 1,000 francs be imposed; in case of a repetition of the offense, confiscation and a fine of 5,000 francs; and should it occur a third time, beside the confiscation, the contract will be relet at the expense of the contractor's surety.

ART. 20. The gravel employed in mixing the mastic must be taken from the beds of streams; it must be thoroughly cleansed of all foreign matter, well washed, free from argillaceous matter, and must be passed through a screen, the holes of which do not exceed 0.2 inch.

ART. 21. The sand for making the mortar for beton and for forming the bed for asphalte, even that employed for bedding the street pavements, the gutters or curbs, shall be sharp, hard to the touch, and will be procured from the best open or working pits, or from the bed of running water.

ART. 22. The cement will be the best from the best known quarries, such as Stefanesti or Rodeni.

ART. 23. The stone used in mixing the beton will be broken stone from the quarries of Paun or Barnova. Each piece must pass through a ring of 2.4 inches diameter; it shall be free from all foreign matter and well washed.

ART. 24. The granite shall have the following qualities :

(*a.*) It will be obtained from the hardest seams of the quarries; it shall be homogeneous, sonorous to the blow of a hammer, without flaws or fractures and free from foreign matter.

(*b.*) It shall have a specific gravity of 2.65.

(*c.*) After an immersion of 24 hours in water, it must not absorb more than $\frac{1}{500}$ of its volume.

(*d.*) When struck a hard blow, it must break in large fragments without leaving any detritus.

ART. 25. The sandstone shall be provided from the hardest beds of the quarries in Roumania. The qualities of the sandstone shall be as follows:

(*a.*) Under the blow of a hammer the sound of the stone must be limpid and pure; a dull sound would indicate interior fissures, and would be sufficient to reject it.

(*b.*) The specific gravity shall be 2.5.

(*c.*) After an immersion of 24 hours in water, it should not absorb more than $\frac{1}{70}$ of its volume.

ART. 26. The limestone for the beton shall be of the hardest quality, chosen from the hardest beds.

ART. 27. The mortar for the beton of the sidewalks shall be composed of two parts of sand to one part of cement in volume. It shall all be mixed and wet with only as much water as may be absolutely necessary. Mortar which may set before being put in place will be rejected. The composition of the mortar for the beton which is to be used on the streets remains to be described hereafter, as also the compressed asphalte.

ART. 28. The beton will be composed of three parts of broken stone to two parts of mortar. The beton which is not used after it is prepared will not be accepted.

The streets which are to be covered with compressed asphalte shall be constructed in the following manner :

(*a.*) The levels will be corrected, the ground shall be shaped to the form of the transverse profile fixed by the city, and will be sprinkled and rolled until it presents a smooth, hard surface.

(*b.*) The curbstones shall then be set, care being taken to cut the joints on the ends as well as on the front edge.

(*c.*) The part corresponding to the gutters will then be made of a bed of sand 2.4 inches in depth, in which will be laid the stones which form the gutters ; the sides of the gutters next to the bed of asphalte will in every case have a range of granite blocks. The remainder of the width of the gutters shall be paved with cobble stones, well assorted, in such a manner that the stones will all be of the same size, or with new sandstone blocks, agreeably to the city authorities. The gutters thus laid shall be well rammed with a rammer, and over all shall be spread a layer of sand 0.4 inch deep.

(*d.*) That portion of the street which is to be covered with asphalte, will receive a bed of beton, the depth of which, after pilonnage, shall not be less than 6 inches. Upon the beton shall be spread a layer of brick broken to 0.4 inch in size.

(*e.*) On this shall be laid the compressed asphalte, when the foundation shall have reached the desired consistency and become well dried.

(*f.*) The asphalte will be laid according to the methods usually adopted on works of a similar nature ; the method of doing this remains to be discussed at a future day by the city and the contractor.

Art. 33. The surface of the sidewalks shall be done over when needed, sprinkled and rolled ; over this shall be laid a bed of beton which after compression shall not have a less thickness than 3 inches. On this shall be spread the mastic to a depth of 0.8 inch. This mastic shall be composed of asphaltic mastic, gravel and bitumen in the following proportions : Asphaltic mastic 100 parts, gravel 60 to 72 parts, and bitumen 6 to 10 parts. The mastic and bitumen shall be melted in a portable kettle, so that they may be carried to the spot where they are to be used. The mastic shall be broken into small pieces, and shall not be poured until it is entirely melted. During the melting the mastic shall be kept stirred, so that a thorough mixture may be insured and the mastic not allowed to burn. The mastic thus prepared shall be run over the beton in such a manner as to spread evenly over it. Before the complete solidification of the mastic which forms the surface, sand will be spread over it and fixed by light ramming. At the junction between a cold and a hot bed, the edge of the cold one shall be reheated by spreading over it a coating of melted bitumen that will then be removed and the final bed laid.

Art. 34. The tempering of the mastic should be such that it will support, at a temperature of 77 Fa., the point of a rectangular pyramid of a height equal to one side of the base, without a depression occurring greater than 0.2 inch under a pressure of 154 pounds continued for five minutes.

Art. 35. The sidewalks at carriage doorways shall have a foundation of beton 4 inches in depth after ramming, and the bed of mastic will be 1.2 inches.

Art. 37. The junction of the mastic with the walls of the buildings bordering the streets will be made by means of a skirting, that is to say, the plaster will be scraped from the walls of these houses to a height of 2 inches above the level of the sidewalks, and the asphalte will be plastered on the bricks to the thickness of the mortar.

The junction of the stratum of mastic with the posts, curbs, hydrants, and other objects of various natures will be made by heating these objects by a coating of melted asphalte that will be then removed and the final bed laid.

Art. 42. The blocks and the cobble stones that the city authorities may declare unfit for using again will be broken and used as Macadam on streets designated by the city ; they shall be constructed as follows :

(*a.*) The ground shall be well graded, giving it the form and inclination of similar streets.

(*b.*) On the surface thus prepared there shall be laid first, gutters of cobble stones to a width of from 2.5 to 3.3 feet. These shall be laid as specified for streets paved with cobble stones.

(*c.*) On that portion of the ground designed for the roadway there shall be spread a bed of broken stone of 6 to 9 inches, reducing, after wetting and ramming, to 4 to 6 inches in depth.

(*d.*) On this strata of stone there shall be spread a bed of sand, which shall be well wet and rolled, so as to obtain a smooth and uniform surface.

Art. 49. * * * * From the provisional reception, the maintenance of the work completed will be performed by the contractor till April 30th, 1830.

Art. 50. All unexpected degradations, all badly executed work, not only during the progress of construction, but also during the years of maintenance, will be repaired by the contractor, whatever may be the cause of the degradation, and with the least delay, without waiting to be notified by the city, under penalty of a fine ot 50 francs for each day of delay after the three first days which follow that of the discovery of the degradation.

Besides that penalty, if the contractor does not proceed immediately to the reparation of these unforeseen accidents, the city shall have the authority to put them in good repair in his place and at his expense, by deducting the cost from the estimate.

ART. 51. On April 1st, 1878, the city will make a new and minute inspection of all the work done during the five seasons. They will compare the state of the work done with that of each of the estimates made at the end of each season, and if it be found that the works are in compliance with this contract, and in good condition, they will finally accept them and make out a new estimate, of which a duly certified copy will be furnished the contractor.

ART. 52. On May 1st, 1880, the contractor having satisfied the first part of the obligations devolving upon him under this contract, binds himself under a new obligation to the city, that of maintaining for a period of fifteen years, from May 1st, 1880, to April 30th, 1895, the works executed, in consideration of a price agreed upon in advance.

ART. 53. The maintenance consists in repairs, renewals and furnishing materials necessary to the sidewalks, curbs, gutters, etc.; in doing all kinds of work and furnishing materials necessary to at all times maintain the surface of the streets and sidewalks paved by the contractor, in a perfect state of uniformity.

The uniformity of the surface of the streets and sidewalks shall be determined by the use of a templet of iron formed to the normal curve adopted for the surface of the streets and sidewalks. This templet, applied to a street or sidewalk, must not present at any point a swell or depression greater than 0.4 inch.

The surface of the streets and sidewalks shall not show any cracks. The connections with the curbs must be perfect.

ART. 54. Whenever the curbs, the sidewalks, the roadways, the water ways—in a word, everything that goes to make a part of the street, becomes subject to a displacement or derangement from any cause whatever, the contractor will be obliged to repair them immediately, in conformity to the preceding article. Exception is made, however, to all that applies to the construction or repairs of water or gas mains, which will always be at the expense of the respective grantees.

ART. 55. The contractor shall proceed at once, at his own cost, to repair any degradation whatever, without notification from the city.

Whenever a notification from the city becomes necessary to warn him to proceed with the reparation, the contractor will be liable to a fine of ten francs, and he will be obliged to proceed with the reparation within 24 hours after the receipt of such notification.

If, after that notice, the contractor fails to proceed with the repairs of the degradation indicated by the city, the latter shall proceed with the work ; it shall make the reparation, submit the contractor to a fine of fifty francs, and retain the cost of the reparation and fine from any sums that may accrue for the maintenance of the streets, and in case that should not be sufficient, from the sureties.

In case of a repetition of this offense occurring during the course of a single year, the city, besides the right which it always reserves to make the reparation on the account of the contractor, in conformity with the preceding paragraph, will impose on the contractor a fine of 1,000 francs.

Finally, on a third infraction in the course of the same year, the city has the right to make besides the reparations after the rules established above, and of re-letting the work of maintenance to another at the cost of the contractor ; if this sale results in a loss to the city, it shall have the right to reimburse itself from the sureties. In case, on the contrary, it results in a profit, the contractor shall have no right to demand it.

ART. 59. The prices for the works executed in conformity to the requirements of this contract shall be :

(*a.*) For compressed asphalte, $5.38 per square yard.

(*b.*) For sidewalk asphalted, $2.81 per square yard.

(*c.*) For roadway, with gutters, paved with sandstone blocks taken from streets now paved, $1.94 per square yard.

(*d.*) For roadway or gutters paved with new sandstone blocks furnished by the contractor, $8.41 per square yard.

(*e.*) For roadway, the gutters included, paved with cobble stones taken from streets now paved in that manner, also for gutters constructed with cobble stones on Macadamized streets, $1.02 per square yard.

(*f.*) For roadway paved with cobble stones furnished by the contract or, also for gutters paved in this manner on streets covered with compressed asphalte, $1.91 per square yard.

(*g.*) For Macadam roadway, $1.45 per square yard.

(*h.*) For granite curb to asphalted sidewalks, $0.95 per lineal foot.

(*i.*) For sandstone curb to asphalted sidewalks, $0.67 per lineal foot.

(*k.*) For curb cut from the old sandstone slabs or flags, $0.39 per lineal foot.

(*l.*) For granite curb used in the edging for asphalte of streets, $0.81 per lineal foot.

ART. 57. The price for maintaining the roadway and sidewalks during 15 years from May 1st, 1880, to April 30th, 1895, shall be calculated from the total area constructed by the contractor, and shall be as follows :

(*a.*) For maintaining compressed asphalte streets with granite curb, $0.09 per square yard per annum.

(*b.*) For maintaining asphalte sidewalk, whatever the kind of curb, $0.04½ per square yard per annum.

(*c.*) For maintaining roadway or gutter paved with old or new stone blocks, $0.29 per square yard per annum.

(*d.*) For maintaining roadway or gutters paved with cobble stones, $0.09 per square yard per annum.

(*e.*) For maintaining Macadam, including the sidewalks and curbs, $0.15 per square yard per annum.

ART. 58. Payment for work done will be made to the contractor in the following manner:

One-quarter of the work done will be paid for in cash, the remainder in City bonds.

The bonds will have 15 years to run ; they will bear interest at 6 per cent. per annum, payable semi-annually ; they will be delivered to the contractor at par.

ART. 67. The sums due for the maintenance of streets and sidewalks during the 15 years which follow the construction—from May 1st, 1880, to April 30th, 1895—will be paid by the City of Jassy, at the end of each month, after deducting all that the contractor owes to the City for repairs made in his name and at his expense, and fines. These payments will be made in cash.

ARTS. 69 to 71, inclusive, provide that the contractor shall place a guaranty of 100,000 francs in bonds of Roumania, which may be replaced by bonds of the City. In addition 10 per cent. of contract price shall be retained, until the sum of 400,000 francs is reached, making a total guaranty of 500,000 francs. This guaranty shall be returned to him as follows : May 1st, 1878, 250,000 francs ; and at the end of the years of maintenance, May 1st, 1895, 250,000 francs.

ART. 72. Six months before the expiration of the fifteen years of maintenance, the City authorities will make a general inspection of all the work done and maintained and make an estimate of it.

ART. 73. If during this inspection the City discovers the necessity of any repairs, the contractor shall do the same at his own expense during the following six months, in such a manner that the streets maintained by him shall be turned over to the City in a good condition on the day the contractor completes his contract.

Should the contractor refuse to make these repairs, the city shall proceed to do it at the expense of his warranty.

ART. 74. At the end of the 15 years of maintenance, if the streets are in a good condition, considering only the effects of the weather, but presenting no degradations, the final acceptance of the works will be made and the guaranty deposited by the contractor returned to him.

The coupons of bonds deposited as guaranties are the property of the contractor, who will have the right to collect the sums corresponding to each. The contractor shall also have the right to replace the bonds deposited by him as guaranty whenever they may become extinguished through the process of drawing lots.

ART. 89. The system allowed for the pavement of the carriage ways is that of compressed asphalte. The manufacture of this compressed asphalte will not depart from the rules actually adopted by science and experience, the mode of execution of the compressed asphalte remains to be agreed upon hereafter between the City and the contractor.

APPENDIX No. 5.

Abstract of Specifications and Schedule of prices for the construction and maintenance of foot-paths and sidewalks, in asphaltic mastic, and the Places and roadways in compressed asphalte belonging to Municipal service of Paris, from January 1st, 1878, to December 31st, 1882. Paul Crochet, Contractor.

ARTICLE 1. The work has for its object—1st. The maintenance and construction of foot paths and sidewalks in asphaltic mastic, situated in the public ways. 2d. The maintenance and construction of compressed asphalte pavements. 3d. The new works ; these works will comprise all the pavements in bitumen and asphalte.

ART. 4.—The present letting is made on a scale of prices, and the amount of the work is completely undefined, so that the contractor cannot make any demand on account of any changes that the expenditure may be subject to.

ART. 6. When the adjoining proprietors or other parties in interest have to bear the cost of the works detailed above, or to contribute in any proportion, the contractor will be held to execute the work at the same price as that done at the cost of the City of Paris, and conformably to the orders of the Engineers.

ART. 7. Foot pavements shall have the widths determined by the administration, they shall be composed of pavement in mastic supported on the side of a public road by a curb the height of which shall commonly be of 6¼ to 4 inches.

ART. 9. The ordinary curbs in granite shall be 11¼ inches wide on top with a total fall across of ⅜ inch ; 13 inches wide at the base, which shall be horizontal ; 11¾ inches high on the front face, which shall have a batter of 1¼ inches.

ART. 16. The mastic pavements will be formed of a layer of pure asphaltic mastic at least ₇⁄₁₆ inch thick, resting on a bed of hydraulic concrete 4 inches thick which comprises a covering of hydraulic mortar at least ¾ inch thick.

ART. 17. The compressed asphalte pavements will consist of an upper layer of compressed asphalte 1¼ to 2¼ inches thick, resting on a foundation of hydraulic lime or cement, concrete 4 to 6 inches thick covered as above with mortar or upon an old Macadam roadway picked over and covered with a thin coat of hydraulic mortar.

ART. 21. The asphaltic mastic employed either for new or repairing old paving shall be composed of naturally impregnated rock with natural bitumen of good quality, coming exclusively from mineral rocks.

The fictitious bitumens extracted by the purification of the heavy oils of schists, the distillation of coal, those so-called fatty bitumens and all other analogous products shall be rigorously proscribed.

The rock employed after being reduced to powder will be melted with a sufficient quantity of purified natural bitumen to form a mastic which, when cold, presents a homologous mass slightly elastic, and which does not soften under a hot sun. This mastic shall be moulded into blocks. There may also be used blocks of bituminous mastic with a base of slates manufactured by the process of M. Sebille.

ART. 22. The contractor shall be bound to employ under the orders of the Engineer upon each public way the bituminous mastic above described.

The mastic shall be formed of a mixture of natural bitumen, in the proportion of one-twelfth of its weight at most, and the calcareous asphalte rocks of Seyssel, Seyssel-Forens, Pyrimont or Volants, of Val de Travers or Lobsan, or others deemed equivalents by the Engineers.

The mastic, having a base of slate of M. Sébille, will be formed of a mixture of bitumen described in Art. 23, following, and of powdered red or blue slate of Ardennes, powdered chalk of Mendon or of Nanterre and of silica from the basin of Paris, in the following proportions, by weight :

```
Refined mineral bitumen.................................... 30
Ground slate...  ...........................................  35
Powdered chalk............................................ 10
Silica, ground and sifted.................................. 25
```

100

ART. 23. The bitumen shall come as much as possible from the washings of bituminous sandstone or the asphaltic rock of Mnestu, and in their default, from the dry pitch of Trinidad, perfectly purified. It ought to be viscid at the ordinary temperature; never brittle or liquid; drawn into threads it should lengthen and only break in very fine points.

ART. 24. The rock employed should be calcareous, soft, with fine grain, texture fairly compact, regularly impregnated with bitumen so as not to show black and white spots; it should be of a brown color; heated to 122 to 140° F. it should soften and break on being torn. Care must be taken for the areas in asphalte to choose only such pieces as are of the most even grain and richest impregnation. The rock of Lobsan, however, should not be employed alone in the asphalte roadways, it ought to be mixed with other rocks less fat in proportions, which will be determined by the Engineers according to the composition of the other rocks. It should contain at least 7 per cent. of bitumen, and at the most 93 per cent. of lime; its change into mastic must not require more than 9 per cent. of bitumen.

ART. 25. The materials entering into the composition of the pavements are the mastics described in Art. 22, pure gravel grit and natural bitumen to assist the melting. These materials ought to be generally employed in the following proportions, by weight:

Foot pavements with a base of asphalte..	Asphaltic mastic...............	100
	Bitumen	6
	Grit.............................	60
Foot pavements with a base of slate......	Asphaltic mastic...............	100
	Bitumen......................	7
	Gravel........................	50

ART. 26. One month before the award of this contract the competitors must deposit at the office of the works in Paris, samples of—1st, A block of the mastic described above; 2d, Specimens of the asphaltic rocks and the natural bitumens they intend to use; 3d, A note indicating the elements of the composition of the mastics and proportions of the various rocks that they intend to employ in the composition of the asphaltic areas.

The blocks and specimens of rocks and bitumen to have the trade-marks of the works from whence they came and the signatures of the competitors.

The necessary certificates to compete for the contract will not be delivered till after the examination and acceptance by the Engineers of the specimens deposited. During all the term of this contract the contractor can only use materials exactly similar to the specimens deposited.

ART. 27. Provides for continuous inspection of the contractor's works or the right to compel the contractor to manufacture the mastic in the depots belonging to the city.

ART. 31. The lime employed is to be hydraulic lime in powder. It must be brought onto the works in sealed bags, marked with the name of the maker. Only the lime and cement designated in the specifications for the construction and repair of sewers will be allowed.

ART. 32. The broken flint must pass through a ring of 2½ inches and be at least ⅝-inch thick. It must be free from earthy matters and washed clean.

ART. 33. The sand shall be dredged from the Seine and well cleansed from all foreign matter; it shall be screened from all grains larger than ¼ inch for the mortars, or ₁₆³ inch for grit for the mastic pavements; the grit for this last purpose shall be perfectly washed and dried before use.

ART. 34. The mortar of hydraulic lime shall be composed of 5 parts of sand and 2 parts of lime, by volume, furnished in powder; the mixture shall be directly reduced to a paste by adding the quantity of water exactly required to reduce it to the consistency of plastic clay.

The cement mortar shall be composed of one part of hydraulic cement of Bourgogne or Portland cement of Boulogne and 3 parts of sand; the sand and cement shall be thoroughly mixed before the addition of any water. All mortar which shall have set shall be rejected.

ART. 35. The beton shall be composed, ordinarily, of two parts in volume of mortar and three of stone. The mixture, made either by the rake or cylinder, must be perfectly uniform.

All beton not used at the time of making shall be rejected.

ART. 36. The bed of beton for the foundation of the sidewalks shall be well rammed and compressed, and must at least commence to set and dry before receiving mastic or asphalte. The beton shall, in addition, be covered with a layer of mortar ⅘ inch thick.

The gravel for foundation shall pass in every direction through a ring 2 inches in diameter. It must be perfectly compressed and sprinkled with lime grout. This foundation shall have commenced to set before the application of the mastic, and shall be covered with a layer of mortar like the beton.

ART. 39. The ground upon which the mastic pavement is to be placed shall always be previously rammed, watered, and crowned with care. When it is thus made solid the contractor shall spread over it the foundation layer, formed according to the orders of the Engineer, either a bed of beton or of sand covered by a layer of mortar, or a bed of sand impregnated with goudron 2¼ inches thick, or any other foundation prescribed by the Engineer.

In all cases the pavement shall not be laid till the foundation has attained the firmness desired, and become quite dry.

The contractor must conform to the following orders for the manufacture of the mastic to be used for pavements.

The mastic shall be prepared and cast in one or more manufactories belonging to the contractor, and which shall always remain open to the inspection of the engineers and their agents.

The contractor shall, besides, establish in the manufacturing depots, both of asphalte and mastic, offices exclusively for the agents of the administration set apart for the inspection of the composition of these materials. These materials shall not be admitted into the works without a carter's delivery note given by the inspector, setting forth that they have been manufactured in accordance with the specifications.

There shall only be allowed in the works blocks of mastic conforming to the samples deposited and accepted before the award, and bearing their trade-mark, or the old mastics from the walks and streets of Paris. All other bituminous matters, resinous or fatty, found in the works by the agents of the administration will subject the contractor to a deduction of $100 for each time.

To assure the execution of these conditions the contractor must not have in any manufactory, under the same penalty, any other blocks than those which should be prepared in his works, and the old mastics taken up.

The use of the old mastic is authorized in the works of the city in the proportion of one-half with the new. The pieces of the old sidewalks having been perfectly cleaned with great care, and regenerated by the addition of new purified bitumen and a sufficient quantity of powdered asphalte to render the old mastic, when melted, of the aspect and consistence of the blocks in fusion.

This mastic shall be melted in hermetically closed boilers, on wheels of a model approved by the administration, and arranged so that the material can be conveyed from the factory to the place to be used, ready to be employed.

For melting, the mastic is broken into pieces 4 inches cube, then the bitumen is melted and the mastic added little by little.

The grit must not be thrown into the boiler till the mastic is completely dissolved.

During the whole time of the operation the matter must be stirred up almost constantly, so that the combination shall be well made and the mastic not burned.

The mastic being well melted and perfectly homogeneous, it shall be run out in bands of about 5 feet wide, spread with a wooden float, and leveled with a strike, so as to present neither fissure nor joint. The mastic must be perfectly level, and match exactly with the curbs, &c., against which it is laid. For this purpose the parts of the curbs, flags, &c., which will be in contact with the bitumen shall be previously warmed and goudroned.

ART. 40. Upon the soil, well shaped and rammed, shall be placed a bed of concrete, covered with a layer of mortar.

The asphaltic rock, conforming to Article 24, broken down or decrepitated by heat, shall be raised to a uniform temperature of from 248° to 260° F., and carried to the place of employment in vehicles that will prevent as much as possible the loss of heat. It must be completely freed from the water it contains. The use of old compressed, taken from old roads, is authorized for mixture with new asphalte, in the proportion of one quarter of old compressed to three-quarters of new rock, provided that the old shall be cleansed with great care before grinding and mixing with the new.

Asphalte shall not be put on the concrete foundation until it is perfectly set and dry.

The powder shall be spread with a thickness about two-fifths more than the finished thickness, leveled with great care, shall be rammed at first carefully, then gradually augmenting the force by means of cast-iron pilons, heated to the proper temperature in portable furnaces. In specially exceptional cases, the compression may also, with the written permission of the Engineer be accomplished by means of rollers.

In every case, after the pilonnage is finished, the surface shall be smoothed by means of a heat-d iron (lissoir).

The road shall not be open to traffic until it is quite cool.

ART. 41. The specifications referring to the construction of roads and footpaths are applicable to the maintenance of the same. The contractor will be entitled to the old material, and will make the repairs in new material or in the mixture specified in Articles 39 and 40.

ART. 43. In conformity with the contract price, stipulated hereafter, diminished by the rebate of the awarded contract, the contractor must make the necessary repairs to all asphaltic mastic footpaths and areas, furnishing the necessary labor and materials, so that they shall be kept in a proper state. He must each year of the duration of the contract completely relay, in new material, at least the fifteenth part of the surfaces of mastic and compressed asphalte. The surfaces in mastic must be properly plane and regular, presenting neither hollows nor projections of more than three-eighths of an inch in a circle whose radius is $3\frac{1}{4}$ feet. These surfaces must be free from fissures.

ART. 45. As the works in asphalte or mastic are received by the engineers they will pass into the charge of the contractor who will receive for the maintenance the price stipulated, commencing from the 1st of January next following their acceptance, whatever may be the date of said acceptance.

ART. 46., The contract prices diminished by the rebate of the award are applicable to the entire surface occupied by footpaths or compressed asphalte, whatever may be their condition.

In the nine last months of the year installments may be paid on the contract when the engineers recognize that the conditions have been loyally carried out. The accumulated sums of these installments must not exceed four-fifths of the amount of the sums which shall be due after the time has expired. The balance of the contract price of the year will be paid in the course of the first quarter of the following year.

ART. 47. The repairs over trenches for sewers, water and gas-pipes, or other works, will be paid for once at the schedule price, but no demand for further payments on account of sinkings or other dilapidations will be entertained, and the surfaces on these trenches must be kept in the same good condition as the others. For the purpose of securing settlement, the contractor may keep the trenches repaired with flint (Macadam) not longer than 15 days.

ART. 49. All damages in the bituminous surface, such as fissures or cracks of at least $\frac{1}{10}$ inch in width, or parting from the curbs $\frac{2}{10}$ inch in width, any lifting up or breaking away of the mastic for at least $\frac{2}{10}$ in depth, depression in consequence of settlement of $\frac{4}{8}$ inch at least in depth under the straight edge, $8\frac{1}{4}$ feet long, will subject the contractor to a deduction of 8 francs (38 cents) per day, when the repairs shall not have been done within 48 hours after notice given by the Engineer.

ART. 51. During the continuance of frost, and during the first month after the commencement of the thaw, there shall be no repairs to the pavements maintained by the contractor, and the inspection for defects shall be suspended, but the contractor shall fill with sand and gravel any holes in these pavements within 24 hours after notification by the Engineer, under a penalty of 10 francs ($1.93) for each day they remain unfilled. He may be authorized, in exceptional cases, to fill the holes with broken flint or melted bitumen, but must replace the flint or bitumen with asphalte as soon as the weather permits. It must be so arranged that the main repairs, intended to re-establish the normal outline of the roadways, are effected from May 1st to November 1st.

ART. 57. The contractor shall execute in private houses the junctions rendered necessary by changes in the public way, which will be paid for according to the price of his contract, subject to the rebate when the works are executed on account of the city.

ART. 65. When a workman leaves one of the districts of the works under the Municipal service, he must have a certificate from the contractor showing the cause for which he left.

This certificate shall be submitted at once to the Engineer, who shall be at liberty to refuse the right of employing the said workman, without the contractor deriving therefrom any excuse for not furnishing. when requisite, the number of workmen required. In default of a certificate, the workman cannot be admitted, except on the written order of the Engineer.

NOTE.—There are 75 articles in this contract, those not given referring to the setting of curbs, etc., transport of materials, and the relations between the Engineer and contractor.

SCHEDULE OF THE PRICES FOR THE WORK SPECIFIED ABOVE.

NOTE, —All the prices below comprise the incidental expenses and the profits of the contractor and are subject to the rebate of the award.

(This contract is let in three lots, the rebates are 7%, 20.2%, and 14%, respectively.)

DAY WORK.—The day of a workman, cart or machine shall be ten hours of effective work in all seasons ; fractions more or less shall be counted by the hour. or $\frac{1}{10}$ of the day.

The night hours shall be paid half as much more as those of the day, excepting watchmen. Night hours will be counted only from 7 P. M. to 5 A. M. in summer, and from 5 P. M. to 7 A. M. in winter. The summer period begins March 1st, and winter November 1st.

1. A day of a laborer	.96½
2. " an ordinary mason	1.06
8. " an asphalte helper, or of a mason's or paver's helper	.87
4. " granite cutter	1.35
5. " sandstone cutters, pavers and asphalte workers	1.26½
6. " watchman	.53
7. Night watchman	.79
8. Day of one-horse wagon and driver	2.70
9. " two-horse " "	4.05
21. 1 cubic yard of stone broken for concrete	1.22
22. " ground hydraulic cement	5 13
23. " river sand	1.18
24. " pit sand	.89
25. " river sand, washed and dried for mastic	1.48
27. 100 pounds of Roman cement	.58
28. " Portland cement	.67
29. " asphaltic rock	.68
80. " mineral goudron, from Lobsang, Bastennes, or other recognized as equivalent to them, and purified Trinidad or Maestu	3.24
49. 1 cubic yard of mortar, composed of two parts of ground hydraulic lime and five parts of sand	3.10
50. 1 cubic yard of mortar, composed of one part of hydraulic cement of Bourgoyne and three parts of sand	5.02
51. 1 cubic yard of mortar, composed of one part of Portland cement and three parts of sand	6.42
52. 1 cubic yard of concrete, composed of three parts of broken stone and two parts of mortar (No. 49)	8.03
58. 1 cubic yard of concrete, composed of three parts of stone and two parts of mortar (No. 50)	3.99
54. 1 cubic yard of concrete, with mortar (No. 51)	4.65
55. 100 pounds of natural bituminous mastic in blocks, made from rock of Seyssel, or other equivalent, ready to be employed	1.02
60. 100 pounds of compressed asphalte, taken from streets to be repaired, shall be taken by the contractor (without rebate) at	.35
61. 1 square yard of old sidewalks in mastic shall be taken by the contractor, without regard to its thickness and without rebate, at	.19
70. Taking up 1 square yard of compressed asphalte, piling the material included	.014
71. Taking up 1 square yard of mastic sidewalk, with piling the material	.008
78. Cleaning and leveling an old Macadam road to secure a surface of mortar for compressed asphalte, per square yard	.10
93. 1 square yard of new natural asphaltic mastic, 0.6 inch thick	.53
94. Greater or less value of each $\frac{1}{100}$ of an inch in thickness	.045
95. 1 square yard of sidewalk relaid in natural asphaltic mastic, 0.6 inch thick, the old material belonging to contractor	.40
96. Greater or less value of each $\frac{1}{100}$ inch in thickness	.025
97. 1 square yard of pavement, 0.6 inch thick, composed of one-half new and one-half new and one-half old mastic	.50
98. 1 square yard of repairs of pavements composed as above	.32
99. For each $\frac{1}{100}$ in thickness, more or less	.02

102. 1 square yard of compressed asphalte, 1.6 inches thick, comprising regulating the surface of the ground, but neither excavation, embankment nor foundation of concrete.. 1 73
103. For each $\frac{1}{10}$ in thickness, more or less... .17
105. 1 square yard repairs of road 1.6 inches thick, the contractor retaining the old material.. 1.04
106. For each $\frac{1}{10}$ in thickness, more or less... .17
108. 1 square yard of foundation for pavement in asphalte or mastic, comprising regulating and ramming the roadbed, but not excavation.
109. 1 square yard concrete of hydraulic cement rammed to 4 inches in thickness, comprising a covering of mortar, No. 49, at least $\frac{1}{10}$ inch thick .... .34
110. 1 square yard concrete of cement, No. 53, rammed to 4 inches thick, comprising a covering of mortar, No. 50 .... .46
111. 1 square yard concrete of cement, No. 54, rammed 4 inches thick, covered with mortar No. 51........ .... .54
112. 1 square yard on natural soil, with a bed of sand 0.8 inch thick .... .04
117. Repairs in which all or a part of the old materials are used, will be paid for at three-quarters of the prices above mentioned.
207. 1 square yard of sidewalk, in natural mastic, half an inch thick, comprising a hydraulic lime concrete foundation, 4 inches thick after ramming, with regulating but not excavating........ .... .93
208. Note. When the foundation is in cement concrete the price above will be increased by the respective differences between each of the prices of Nos. 110 and 111, and that of 109.
209. 1 square yard of mastic like 207, on the natural ground, covered with a bed of sand 0.8 inch thick .... .63
210. 1 square yard of mastic like 207, on the natural ground, covered with hydraulic mortar... .68
211. Note. When the pavement is one-half new mastic with old material, the prices above will be dimished by.. .09
212. 1 square yard of compressed asphalte 1.6 inches thick, including a foundation of hydraulic lime concrete, rammed 4 inches thick, including dressing and ramming the soil but not excavation................... ... . 2.07
213. 1 square yard of compressed, with concrete of cement No. 51........ .... 2.21
Additional price for work executed on embankments more than 1 yard high, or on trenches, whatever may be the thickness of the pavement or concrete.
215. 1 square yard of mastic .. .. .22
216. " " compressed.. .36
217. Maintenance of 1 square yard of sidewalk in asphaltic mastic, in conformity with these specifications, per annum... .05
218. The same for roads in compressed asphalte............................... .19
219. Additional price for maintaining cross-walks of compressed asphalte on Macadamized roads, and of gutters bordering them.................................... .10
249. For works not mentioned in the present schedule, the prices in the schedules now in force for maintenance of public ways or sewers and water service will be paid: which prices will be subject to the rebate of the present letting.

THE CONSTRUCTION AND MAINTENANCE OF ROADS.

By Arthur Spielman and Charles B. Brush, George D. Ansley,
A. B. Hill, Charles Douglas Fox, E. Lavoinne, E. B. Van
Winkle, B. F. Morse, E. S. Chesbrough, E. R. An-
drews, C. Shaler Smith, M. Merriwether,
J. E. Hilgard, D. E. McComb, F. Rinecker,
J. J. R. Croes, John Bogart, C. C.
Martin and Edward P. North.

A. Spielman and Charles B. Brush (Spielman & Brush).—
Some of the views expressed in Mr. North's paper on "the construction
and maintenance of roads," being at variance with the results of the experience of our firm, in the building of 36,000 square yards of Telford roads
in 1875 and 1876, and about 25,000 square yards in 1878 and 1879, part of
this latter amount being now in the course of construction, we herewith
submit the principal facts in relation to these roads, and our conclusions
therefrom.

The roads are 80 feet wide between house lines, and are located in the
northern part of Hudson County, New Jersey ; they are built exclusively of
trap rock, obtained from and along the line of the road, the stone for the
upper courses having been broken by a stone crusher erected on the road
by the contractor, the average haul from the crusher not exceeding 2,000
feet. No binding except the screenings and detritus of the stone was allowed in the work, and in each case the foundation is of rubble, 8 inches
deep, and the superstructure of broken stone, 4 inches thick, when compacted.

The roads built under our direction in 1875–1876 may be divided into
two classes :

1. Roads by the side of horse railroad tracks.
2. Roads free from horse railroad tracks.

In the first case the width of the roadway from curb to curb is 55 feet, which includes 18 feet of trap block pavement for the tracks, and 5 feet for the gutter.

In the second case the width between the curbs is 40 feet, including 5 feet of block pavement for the gutter.

In both cases the 8-inch foundation was first carefully laid, and great care taken to allow for perfect sub-drainage. Cess-pools, filled in with broken stone, were built at intervals of about 200 feet on both sides of the road, which collect all the water that accumulates in the foundation of the pavement, and these cess-pools are drained by 6-inch stoneware pipes into adjoining receiving basins.

A particular illustration of the importance of this sub-drainage came under our notice. In December, 1875, just after considerable of the foundation had been laid, legal difficulties arose, the work was suddenly stopped, and remained in this unfinished condition until the spring of 1876, receiving in the meantime the wash from the adjoining hill-sides. When the work was recommenced the interstices between the foundation stones of the pavement in many places were filled in with earth. After unsuccessfully attempting to remove this earth, the foundation at these points, as far as they could be ascertained, was taken up and relaid : but as soon as the superstructure of the road was completed we found that, in certain spots, it was always wet, and the surface of the road was continually broken. These spots invariably indicated the points where the foundation was clogged, and the difficulty was only effectually remedied by relaying the foundation, or by building blind drains which carry off the accumulating water.

On the top of the foundation thus prepared, 2-inch stone was then put on, sprinkled, and rolled with a horse roller of 150 pounds per inch run. The one-inch stone and screenings were then spread, sprinkled, and rolled with a steam roller of about 400 pounds per inch run.

After the rolling was partially completed the passing traffic was allowed upon it, and any large stones that came to the surface, as well as all small stones that failed to bind, were raked off and sent back to the crusher to be re-broken for screenings. No water-worn or other rounded stones were allowed in the work. Advantage was taken of every rainfall to roll the surface of the road, because we found that it could be compacted much more thoroughly in wet than in dry weather. Where the pavement was laid in a soft substratum it required nearly double the amount of rolling sufficient for a solid foundation.

The grades of these roads vary from 6½ feet to 8 inches per 100 feet and the crowns vary from 12 inches to 8 inches.

The roads have now been open for traffic about three years; those along the railroad tracks are used by about 600 wagons per day, the others by about 400 wagons per day. Fully one half of the traffic on both roads consists of heavy beer wagons from the adjoining breweries, stone trucks and ice carts, ranging in weight from 3 to 6 tons; the balance of the traffic, of ordinary farm wagons, carriages, etc.

No especial care has been taken of the roads, except to see that the gutters and culverts are kept clean. No ruts have ever appeared, and the surface is now smooth and in good condition. In hot weather the roads are somewhat dusty, and in long dry spells they will loosen and break up in spots, where disturbed by the corked shoes of horses drawing very heavy loads, but after the first rain the surface immediately rebinds and again becomes perfectly smooth.

Wind and water are, perhaps, the two greatest enemies of Macadam roads; the wind, by blowing off the slight dust which naturally accumulates on the surface, removes from the road the cushion, which is not only a relief to the traveler, but which also preserves the metal of the road from a vast amount of wear and tear; the water, by flooding the road, has sometimes the same effect as the wind, and if by any means the surface of the road is exposed to a running stream the stones are sure to loosen.

The only effectual remedy we found was to raise the crown of the road sufficiently to shed the water quickly into the gutters, and to keep the road sprinkled, so that when the winds and floods came the surface would be smooth and compact and not liable to their disintegrating influences.

On the roads by the side of the horse railroad tracks the wear has been about an inch and a half during these three years, while on the roads free from these tracks the wear has been about one inch on the crown and perhaps a half inch on the sides.

The cash cost of the Telford pavement laid under our direction in 1875–76 was ninety cents per square yard.

The stone was broken by a ten-inch "Blake" stone crusher at the rate of about twenty cubic yards in ten hours. The size of the stones as they came from the crusher was: 50 per cent., 2 inches size; 25 per cent., 1¼ to 1 inch size, 25 per cent., screenings and pea dust.

The cost of the crusher, engine, boiler, &c., set up complete, was about $2,500.

The cost of working per day, independent of the original cost of the machinery and interest thereon, and also independent of any royalty on the stone, was found by the contractor to be as follows :

Repairs, lubricants, wear and tear on crusher and engine, about.......................... $5.00
1 Engineer, $2.50 ; 1 feeder, $1.50 ; 1 screener, $1.50 ; 5 laborers quarrying and breaki. g
   up stones at $1.00...................   .... 10.50
1 team hauling stone................   .... 5.00
1 Coal half ton 2.50

    Cost of preparing and crushing 20 cubic yards of stone...........................$24.00
    Cost of 1 cubic yard, $1.20.

The roads built under our direction in 1878, and now building, are the same in every particular as those built in 1875-76, except that they occupy only 20 feet in width of the crown of the road; the steam roller was not used, and as it is very difficult to obtain an abundant supply of water in the locality, we have to rely on the rainfall for sprinkling, and do all our rolling in wet weather. A horse roller is used of 150 pounds per inch run.

Some 3,000 square yards of these roads were completed one year ago, and have been, since that time, subjected to a daily traffic of about 150 wagons, principally carts, loaded with stone and dirt. The surface of the pavement is now as nearly perfect as it is possible to imagine that of a macadamized road to be.

The cash cost of these roads is eighty cents per square yard.

CONCLUSIONS.

A Telford road may be practically divided into two parts.

1. The foundation, which should be uniformly secure, and which should be at the same time a perfect blind drain.

2. The superstructure, which should be a durable, water-tight roof.

If these conditions are complied with—if proper materials are used in the construction of the road, and reasonable attention is given to its maintenance, the result will be as has been claimed, a durable road, unsurpassed for comfort of travel, and one to be preferred to all others for sanitary reasons.

If these conditions are not attended to the road will last but a short time.

The foundation is of the first importance. It should be eight inches in depth. More than this is a waste of material, and a less depth is not sufficiently secure for want of proper bond. It must be laid as close as possible by hand, then the interstices at the top wedged and sledged, until the small stones that compose the superstructure cannot work down, and fill the interstices at the bottom of the foundation.

Too much emphasis cannot be given to this part of the work. It is not only essential to perfect sub-drainage; it is equally important in the great saving of the cost of building the road. A loose foundation, which allows the small stones to settle down upon the large ones, will require nearly as many again of the small stones before a proper surface can be obtained, hence the cost of the superstructure will be nearly doubled. Of course, if the small stones work down among the large stones, the latter will work up the surface, and ultimately ruin the road.

An excellent test of a foundation, when the substratum is firm, is to drive a loaded truck—weighing about three tons—over the pavement before any of the upper courses are placed thereon; if the foundation has been properly laid no ruts or other displacement will occur.

As to the size of the foundation stones, we prefer them large rather than small. Nothing is so dangerous as thin slabs. A large stone now and then, say ten inches wide, well bedded, seems to act as an anchor for the rest, and we have yet to find an instance where the small stones have broken loose from such a foundation stone.

In regard to the superstructure, we are convinced that if the material is crushed trap, any increase over four inches in depth is a waste of material. We found this to be the most expensive portion of the road, the cost of the eight-inch foundation being to the cost of the four-inch superstructure as one is to two. Four inches will answer all requirements as well as any greater depth, because after the metal has worn down two inches, the road, owing to unequal wear, will need to have a new coating in any event, and the amount saved in the first cost and the interest thereon, by making the superstructure only four inches deep, will keep the road in repair for many years.

In regard to steam rolling, it is often questionable whether it is essential, or even desirable, in the building of Macadamized roads, especially when the road is built of New Jersey trap rock.

The principal action of the steam roller is to crush the stone into the crevices, and the result is, that a crust is quickly formed. On the other hand, the horse roller rattles around and shakes the small stones about, until they are firmly bedded upon the rough but firm foundation and upon each other. No crust is formed, but, on the contrary, a compact homogeneous mass, which result is much more to be desired. Again, the road bed upon which the pavement is laid often varies very greatly; frequently a rocky bottom adjoins a soft stratum; on one side of the road may be an excavation and the other side a fill. In such a case, a heavy roller is much more likely to disturb the uniformity of the foundation than a lighter one, no matter how great care may have been taken to provide for emergency.

If it be necessary to finish the surface of the pavement within a week or two, a steam roller must certainly be used, but we believe that rapidly made roads are much less durable than those whose construction extends over a long period of time.

It required at least three months to finally form the surface of the roads built by us in 1875–76, while on the roads now building, some of the sections have required as much as six months. In the meantime, the surface is kept free from loose and rolling stones, so that there is no brutal pulling through the road metal.

After a road has been slowly compacted in this way, we believe the surface will be found much more durable than that of any rapidly made steam rolled road.

As to "binding," our experience has been that during the construction of the road the less foreign material used the better, unless, perhaps, along the edges of pavement which has only an earth support. In such a case, it is necessary to bind the edges as quickly as possible, in order to prevent the sides of the road from spreading while the surface is forming. After that is accomplished, very little wear comes on the extreme edges of the road.

As an aid in the rapid formation of a fine surface, a little yellow clay placed just below the upper course is almost invaluable, but when the crust is broken the danger is that the surface will soon disintegrate, while if, instead of the clay, stone dust is placed between the courses of broken stone, and a top dressing one inch deep of screenings, such as are presented herewith, is spread over the surface and is thoroughly worked into the broken stones, the surface is equally fine and much more durable; if a spot does loosen here and there, it does not spread and a little moisture quickly rebinds the loosened stones.

Finally, as to the stones for the superstructure, we greatly prefer machine-crushed to hand-broken stones.

1. Because they are much more uniform in size, each having actually passed through a revolving screen.

2. Because the edges of the stones are much sharper and bind better.

3. Because from the machine alone can we obtain the screenings and detritus which we consider so essential for compacting the road and for satisfactory top dressing.

GEORGE D. ANSLEY.—My experience is decidedly in favor of steam rolled Macadam or Telford roads over those formed by horse roller; in fact, I have altogether given up the use of the latter, and employ a 15-ton Aveling & Porter, the result being far greater economy in the end as to outlay, and a decidedly smoother and more permanent surface is obtained.

As to compacting with traffic, I am altogether opposed to it, as being inhuman toward horses and extravagant in the waste of material. I speak of the case when any considerable extent of roadway is to be covered; but in small repairs, or what is technically called "darning," I first pick up the margins of the depressions to the depth of an inch or two, and then flush up with stone broken to pass through a 2-inch ring; the edges of the patch are then covered with road grit, obtained at hand and pounded with a rammer. In these repairs it is found that the horses feet avoid the fresh stone, while the wheels of the vehicles run over the patch and compact it gradually from the edges to the centre and a very good "mend" is thus made.

In reading over Mr. North's valuable collection of short histories of road making, I was particularly attracted by the mention of ramming, on page 103. There is a short mention of a rammer 8 inches diameter, weighing 70 pounds. Although steam rollers are far more satisfactory than horse rollers, it seems to me that a s'ill further improvement may be made by the more general introduction of the rammer. In all cases of roads, whether Telford or Macadam, or stone paving, or wood paving, the first imperfections are the same ; the surface may not be worn away materially, but there are depressions or concavities that hold water. Our block stone pavements get into bad order chiefly through unevenness, and a heavy expense is incurred in repairs, while the hollow parts are found to be hard and well set, and the blocks not worn, perhaps, any more than those forming the better parts of the road. A wooden pavement on one of our streets was condemned for being in hills and hollows ; although incidentally there were rotten blocks that broomed and wore away, it was found, on taking it up, that the foundation was uneven, and this notwithstanding that it had been steam-rolled before the blocks were laid on, five years before. All this tells the same tale—the earthy foundation is of unequal density.

Much more attention has been paid to the coating of stone or other material than to the lowest or earth foundation. When a new road is made, the proper form may be given to the earth foundation, and it may be steam rolled ; inequalities then showing themselves may be flushed up and re-rolled, but a roller will bridge over smaller soft places which still remain unseen until the road is completed and heavy traffic put upon it, and then we have saucer-like dips in its surface, to be repaired within a short time after the road is made ; and although the specification may require that these repairs shall be done by the contractor, the surface being thus broken, the road is never so good afterward as it would be if undisturbed.

How the rammer which Mr. North refers to was used is not mentioned ; but if by steam power, it might be neither expensive nor slow.

I am inclined to think that the regular stroke of a rammer is the only method of producing equal densities for roads, as well as for other purposes, and that the most important part of road making for its application is the earthy bottom.

After the ramming to equal density, I cannot see that Telford's method is better than Macadam's for general adoption ; local circumstances, however, would decide in each case.

After reading through all the various descriptions, and adding my own views from observation and experience, I consider the object to be aimed at is as near as possible, a solid bed of stone. This certainly cannot be accomplished by putting on any earthy matter as binding. Broken stone in thin layers, 3 to 4 inches, chinked with fine chippings or screenings until full, and then watered and rolled with steam roller, will come very near the desideratum. Of course the lower strata may be of stone, less hard than the top. The size of broken stone for the upper part is important. A 2½-inch ring allows stone of considerable size to be mixed with lesser ones, and it is these larger ones that first get loose and move about on the surface. For the surface coating I prefer hand broken stone over machine broken, as the form is generally more cubical and less apt to become disintegrated.

Gravel is much more difficult to reduce to solidity than broken stone; but where there is a large supply, and cheap, it is well to follow the principle before mentioned. The gravel should be screened, and the coarser sort laid on first and then chinked up with the finer. If this is done, and all coarse gravel kept away from the top coating, water and steam rolling will make a good road in almost any case without earthy binding. Earth matter works into mud, and should be avoided, unless the gravel is so round and movable that nothing else will keep it quiet.

Where old paving is taken up, and it is proposed to put down Macadam, I consider that there is the same necessity for testing the density of the foundation, and rendering it equal by ramming.

A. B. HILL.—In New Haven we have tried several plans in regard to the binding material of the Telford pavement. Using an inch of loam on the crushed stone, with two inches of screenings over that ; also using sand instead of the loam ; but the best results with us are obtained by using the trap rock screenings alone, spread on in thin layers, sprinkled and roughly rolled. This makes a very solid, firm surface, which does not wear into ruts as soon as the pavements are top-dressed by the other methods. The roller used is the 15-ton Aveling & Porter.

As the grades in New Haven are generally very light, and it is desirable to secure a uniform, smooth gutter, the latter is made of blue stone, 12 inches wide and not less than 4 inches thick, bedded in sand next to the curb, closely jointed, well rammed, and, after the pavement is complete, thoroughly grouted.

A space of three feet outside the gutter stone, between the rails of the horse railroad tracks, and for three or four feet outside the rails, is laid with stone blocks.

The Telford pavement is 16 inches thick at the centre and 14 at the sides, made up of 4 courses of trap rock ; the first or bottom course 7 inches thick at the centre and 5 inches at the sides ; the stones of the size and placed as usually specified for Telford foundation ; the second course, 3 inches thick, of stone simply raked out and sorted at the foot of the trap dikes (not " broken " or " crushed "), varying in largest dimensions from 1 inch to 4 inches, spread on the first course and rolled until solid ; the third course, 4 inches thick of " crushed " stone, also rolled ; the fourth course, or top-dressing, about 2 inches thick of screenings, spread on in three layers, each layer sprinkled and thoroughly rolled in.

The average cost of the Telford pavement in New Haven, including the Belgian blocks, blue stone gutters, crosswalks, inspecting, rolling, &c., was, for 1876, $1.18 per square yard.

" 1877, 1.05   "   "

" 1878, 1.15   "   "

E. LAVOINNE, Engineer des Ponts et Chaussées (through the Secretary).—The criticism of Mr. North upon the Macadam roads in the City of Paris is to the effect that owing to the method of compacting, sufficient stability is not given to the stones to resist the traffic. Mr. Malo is quoted as sustaining this criticism. Even if the bad results in Macadam pavement in Paris were something like what Mr. Malo describes in his rather sweeping remarks, the fair inference would be, it seems to me, that the system of construction was not the best in that location on account of the heavy traffic.

Macadam roads when introduced in Paris to replace the former pavements were considered by many engineers as a blunder, on account of the cost of their maintenance and other peculiarities. It is certain that no Macadam road, even if constructed under the best conditions, could stand the enormous traffic existing in many streets, which is not occasional, as Mr. North states is the case for some of the Boulevards of New York, but continuous and daily for most of them.

As regards the construction of roads, the illustration of what Mr. North calls the French system, such as he saw applied in the repairs of some streets in Paris, hardly gives an exact idea of the standard system adopted by many French engineers. They generally consider that in a perfect Macadam all stones should bear directly against one another by faces as large as possible, not by edges, and that the interstices, previously reduced to a minimum by rolling, should be filled afterwards with a binding which cannot be affected by atmospheric influence nor give access to moisture. Thus far they agree with Mr. North—but they disagree with him as to what is the best binding.

Instead of screenings or very small stones with the addition of dust and water, they prefer to use sand with a small quantity of chalky dust employed when compacting is at an end. They consider, contrary to Mr. North's theory, that when the stones, whose sizes vary between 1½ and 2½ inches, have been thoroughly packed together by rolling before any addition of binding, so that they move no more under the roller, and a beginning of crushing takes place, then an addition of smaller stones is useless for stability ; if very small stones like screenings were then added, they would be crushed and produce an excess of dust injurious to general stability. Sand, injected by thorough watering between the stones is not liable to that objection, since, filling all the interstices, it tends to equalize the pressures between the stones. The addition of chalky dust diluted by water, at the end of the operation, fills the interstices between the grains of sand making with it a sort of mortar and coating for the surface. In my own experience the best results followed this method; the consolidation of the Macadam was very satisfactory at the end, rolling not being spared before the addition of binding. Loose stones occurred only at a few points.

It may be that the roads examined by Mr. North, in Paris, were constructed in too thick layers and too hastily permit the stones, prior to any addition of binding, to have the required stability, much rolling being necessary for this result. This may be the reason why loose stones were seen. If the stones have not been packed and wedged previously by thorough rolling, we cannot expect binding to make them immediately compact.

The system mentioned as used in New York (St. Nicholas Avenue), in which stones of from 1 to 1½ inches are employed for he top course, will, no doubt, do for light travel (light, not heavy carriages); but such pavement would very likely be destroyed by a heavy traffic, as the small stones would then be rapidly ground and disintegrated. From my own experience and that of many engineers in France, I am fully satisfied that the capital difference between the roads in the old Macadam style without rolling, and those that are rolled, whether with steam or horse power, is the degree of internal wear; as the grinding of the stones by their reciprocal friction or internal disintegration is much more rapid under heavy traffic with the former than the latter. In the first case the proportion of disintegrated material, detritus, to the stone is generally large after a short time; very small in the second if proper care has been given to the work.

Evidently much more consideration should be given to the internal wearing, which is of serious consequence as to cost of maintenance, than to the incomplete consolidation of the road immediately after rolling, which could be remedied by more rolling or made up afterward by the traffic itself and by removing the excess of binding material by sweeping.

In conclusion, it is suggested that for a fair comparison between the different systems of constructing the roads, there should be taken into account both the quantity and quality of the traffic, and also the cost of maintenance under the same conditions for a fixed period. Conclusions might be different if full consideration were given to these points.

E. B. VAN WINKLE.—I would say that I am familiar with the roads Mr. North has been constructing for the past few years; that is, with their present condition, and should like to ask Mr. North if their present condition bears any relation to the amount of rolling he put upon them; for instance, the Southern Boulevard, which I now consider to be the best of these roads, and the one that carries the greatest traffic, did that receive the greatest amount of rolling?

E. P. NORTH.—The Southern Boulevard received 0.859 ton mile per square yard, or 5.177 ton miles per cubic yard, which is more rolling than any other road constructed by me had, and more than any road known to me has had applied to it.

The road has stood the wear very well, though part of it is exposed to a very constant breeze from the Sound, which deprives it of the protection that a layer of dust would afford.

The teams on it were counted from 8 A. M. to 5 P. M., and averaged about 300, from four-horse teams, with six or seven tons to the load, to buggies.

E. B. VAN WINKLE.—Next to the Southern Boulevard I should place the streets constructed by Mr. North in the following order as to degree of excellence, always judging by their present condition :

1st. One Hundred and Thirty-eighth street ; 2d. One Hundred and Sixty-seventh street ; and last, Mott avenue. Please, if possible, state which of these received the greatest amount of rolling, and if there were any difference in the quality and size of the metal and the material used for binding.

E. P. NORTH.—One Hundred and Thirty-eighth street, which has two courses of broken stone, each about 6 inches deep before rolling, received less rolling than the Southern Boulevard ; the surface is satisfactory except in one place, where the bottom was bad and mud worked up through the metal where there are some loose stones. One Hundred and Sixty-seventh street is on a heavy grade, part being at the rate of 11 2-10 feet per 100, and the rest with 8 per 100 for a maximum. The first was rolled with both horse and steam rollers, the steam roller ascending by an easier grade. Some clay hardpan was used here in connection with the screenings, both to increase the adhesion of the roller wheels and facilitate the compacting of the road bed. The roller, a 15-ton Aveling & Porter, old pattern, ascended the grade after the application of the hardpan. On the lighter grades nothing but screenings was used for binding, and the rolling was done entirely by steam. This part of the wheelway wears much better than that portion where hardpan was used. No reliable account was kept of the amount of rolling this street received.

The circumstances under which the wheelway on Mott avenue was constructed are fully detailed in Transactions, Vol. VIII., page 110 (May, 1879). On account of its treacherous bottom it probably received less rolling per square yard that any other road, though it was impossible to keep accounts of the amount of rolling done.

All of these streets are Macadamized with two-inch trap and clean trap screenings, excepting that portion of One Hundred and Sixty-seventh street mentioned above.

B. F. MORSE.—The paving of the Cleveland Viaduct, west of the river, over the arches, is laid with New York Medina sandstone ; the road-way is 42 feet between curbstones. with a double track street railroad in the centre.

The ballast used was of the best quality of bank gravel spread in layers of about five inches in depth. Each layer was sprinkled with water and rolled. The last layer, or that directly underneath the stone, was about two inches deep, and was left without sprinkling or rolling to receive the bed of the paving stones. The surface of the ballasting was finished to the true crown of the roadway.

The pavement is laid with blocks, dressed nearly parallel on top and bottom, sides and ends, laid in courses transversely across the roadway. The courses were from three to four inches thick, and from six to seven inches in depth, and the stone from seven to twelve inches in length The stones were set close together, so that no joint was more than one-half inch open for at least two and one-half inches down from the top surface.

No gravel or sand was placed between or on top of the pavement while it was being laid. After the stones had been set in place in sections of fifty to one hundred feet in length of the street a light top dressing of gravel or sand was spread over the surface and swept into the joints with a steel splint broom. The pavement was then thoroughly sprinkled or flooded with water. Then the pavement was thoroughly rammed two or more times with a paver's rammer weighing about ninety pounds ; then the pavement was again washed or flooded and allowed to dry off.

The joints were then filled to a depth of three to five inches with a concrete composed of Trinidad bitumen and coal-tar cement, distilled at a temperature of not less than 600 degrees Fahrenheit, and mixed in proper proportion, so as not to soften or become brittle under heat or cold, and was poured into the joints of the pavement at a temperature of not less than 300 degrees, and then the whole surface was covered with one-half inch of fine gravel or sand, which completed the work.

The pavement on the fixed iron spans was laid in the following manner : Strips of oak plank, varying in thickness from one and a half to three inches, were secured to the iron floor beams, running longitudinally, to give the proper crown to the roadway. On top of these longitudinal floor beams was laid a layer of two and one-half inch plank, joints well broken and spiked down. On top of this layer of plank there was laid two thicknesses of tarred roofing felt, or paper, laid in hot roofing cement, and the whole covered with one-fourth of an inch of plastic pitch, and over this was laid a layer of inch boards or sheathings, breaking joints with the plank under-neath, and thoroughly spiked down.

The paving is what is usually called "Nicholson," and consists of blocks four inches long and three inches thick laid upon and in rows across the roadway, with a three-fourth-inch strip, one and one-half inches in depth between the rows of blocks, and nailed to the flooring, the blocks breaking joints at least two inches with adjoining row. The space between the rows of blocks was then filled with concrete, composed of one part of hot undistilled gas tar to two parts of pitch, mixed with clean lake sand and fine gravel, applied hot and driven into the joints with an iron blade and heavy rammer until the spaces were even full. The whole surface of the paving and gutters was then coated with a top dressing of coal tar pitch and fine gravel, rolled thoroughly with a heavy hand roller. The best quality of seasoned white oak was used for all the wood parts of the pavement and plank floors.

E. S. CHESBROUGH.—I cannot state the average wear of wood pavements; I can only state that it differs very much with regard to different kinds and in different localities. You can easily see that very much depends upon the faithfulness of doing the work and the material used.

In some cases in Chicago wooden pavements have lasted ten years, and even longer; and in others they have become very rough and uneven in three or four years. I am not able to give the precise average, but of course a great deal depends upon the traffic. In the river tunnels the wooden pavements have worn out in less than two years, and where the wheels were confined very much to the same tracks they make ruts in a short time. In other cases, where the streets are broad and clear, and the traffic is spread over a large space, they have lasted a long time; in some cases ten years. It is impossible to give the rule in regard to that unless you take into account various circumstances.

EDWARD R. ANDREWS.—I would like to ask Mr. Chesbrough whether at Chicago there is any very perceptible wear in wooden pavements until decay sets in?

EDWARD S. CHESBROUGH.—Decidedly; I have seen some worn down more than two inches without any apparent decay.

EDWARD R. ANDREWS.—Mr. North states that a well made Macadam road, constructed with trap rock, is, after an earth road, the pleasantest and safest known. But trap rock or other really good materials for making Macadam roads are not available everywhere, and, at best, Macadam roads are only adapted for pleasure travel in parks or suburban towns, where they can be constantly watered and never allowed to get out of repair. Macadam is not adapted for general use in cities. Under heavy traffic the surface is constantly ground into powder, which rises in dust in the summer, and they are very muddy in the winter. Even in Paris, where the maintenance is most thorough, the streets being continually watered in summer in the manner described by Mr. North, and frequently washed after a day of unusual wear, and scraped by a large army of cantonniers, yet, after heavy rains, the mud is frequently nearly ankle deep, and in very hot weather, during the intervals of watering, or in frosty weather, the air is filled with most penetrating dust. Mr. Flad describes the same state of things in St. Louis ; and in Boston, when, in winter, there is no snow to cover the ground, and, on account of the cold, the streets cannot be watered, the dust is intolerable ; and in summer, where, for economy's sake, watering is neglected, a large part of the material with which the roads are made is blown into the sea.

The compressed asphalte, so common in London and Paris, when constructed as thoroughly as it is in those cities, and as that on Fifth Avenue, in front of the Hotel Brunswick, has been, is a most excellent pavement, but it also demands the most careful maintenance. No dirt should be allowed to accumulate upon it. In frosty or in damp weather coarse sand or fine gravel should be spread over the surface to give a good footing for horses—this is done abroad—and then it is not slippery. It is very quiet, and, in fact, has almost all the qualities needed in a perfect pavement ; but it can only be laid on levels, and is expensive.

Stone Block Pavements are in many parts of the country the cheapest and possibly may be the best where the traffic is very heavy, but it is emphatically the worst pavement for streets of residences or wherever quiet is desirable ; and there is no question but that if the incessant din from the rattling of omnibuses, heavy teams, milk wagons, &c., from which one suffers in large cities paved with stone blocks, could be dispensed with by adopting a quiet pavement, the length of life of citizens would be increased and the general health improved. Such would have been the case long ago in New York, had it not been that the wooden pavements laid during the "Tweed" days were such evident jobs. In London, wooden pavements give entire satisfaction. The earliest were not quite successful, but the defects in construction have been remedied, and now broad areas of heavily worked streets previously paved with stone are being laid with wooden blocks, which are found to wear satisfactorily.

In the West, where stone for pavements cannot be had, wooden blocks are largely used ; but, as wood is cheap and can be replaced without much expense, no sound principles are followed in their construction. In the Eastern States, no one will allow that a wooden pavement can be good except when newly laid, when all agree that it is delightful. There seems to be an unwillingness, even among engineers, to give the subject the attention it deserves. All agree that stone pavements are a curse, and that it would be a blessing if a good substitute could be found, but because wooden pavements, as they have been made here, have not been a success, condemn them as a class.

Mr. North has stated what has been the general practice in laying wooden pavements in this country. Many methods have been tried, but they have almost without exception been "laid with *green* or wet blocks, more or less thoroughly dipped in tar, on a bed of sand, not always well rammed, with or without the interposition of a tarred pine board, with transverse joints from one to one and a half inches wide filled with gravel and coal tar," and I might add, the whole done in a most unworkmanlike manner.

The results are what might have been expected. The careless manner in which the joints have been filled, has left many channels open for the admission of water, which undermines the sand foundation, so that there is an uneven subsidence under the passing wheels, and holes, small at first, but daily growing larger, appear, so that the surface is soon destroyed. The result is but little better when tarred boards are laid under the blocks. This practice of tarring wet sappy boards and blocks seems to be an invention to make them decay as soon as possible. It closes up the cells of the wood, so that the moisture cannot escape; fermentation immediately follows, which quickly destroys the strength of the fibres and reduces them to punk. A pavement, constructed in this manner, would fail, of course. Thoroughly seasoned wood might be benefited by the tarring process, but green wood never.

Observe how differently wooden pavements are constructed in London. Mr. North describes several methods, either of which is vastly superior to any of the patented systems used here. A rigid foundation of bituminous or cement concrete is universal. This costs more than sand, but it is permanent, and will prevent the blocks from sinking under the wheels. English engineers, in discussing pavements, call the foundation the true pavement, the blocks being the wearing surface only. The " Henson " pavement, with some modifications, strongly recommends itself to my mind as the best for this country. Instead of a layer of tarred paper on the concrete, I would use a thin layer of pitch, with oil enough in it to make it permanently slightly plastic, setting the blocks upon it while hot and soft, using the strips of tarred felt between the rows, and driving the blocks together as described by Mr. North. The tarred felt would make a very close joint. Then pour melted pitch over the whole surface, taking care to fill every crevice, and upon this spread fine sharp gravel, which will work into the ends of the blocks and form a surface resembling macadam, and afford a far better footing than wide spaces between the rows, which serve as receptacles for mud and dust. It is easy to keep this pavement clean. No water can penetrate it, so that it will not be injured by frost. The blocks themselves, if creosoted, will not absorb water, and if laid without spaces between the blocks, the drainage will be surface drainage solely, which is of the first importance.

But the pavement would be short-lived, if *green* and wet blocks are used. It is not practicable to use, as Mr. North says is the case in London, "wood better seasoned than the pine generally used by house carpenters in this country." Seasoned wood cannot be obtained in sufficient quantities here. But, what is far better, it can be preserved from decay, I have no faith in any method of wood preservation for paving blocks which does not exclude water. The blocks are so short, that any soluble preparation is quickly washed out of them, and, if not made waterproof, they are certain to absorb the seeds of destruction from the filth in the streets. The blocks should be well saturated with creosote oil, whose chemical constituents act preservatively upon the. fibres of the wood, by coagulating the albumen of the sap, while the fatty matters act mechanically in obstructing the pores of the wood and keep the water out. At the same time, as oil cannot be injected into wood full of moisture, the thorough artificial seasoning, which forms a part of the process of creosoting as carried on in this country, is as useful to the timber as any of the metallic salt processes.

By thoroughly creosoting the blocks, expansion and consequent throwing out of the blocks is prevented. They will not shrink or expand. The wood is also rendered homogeneous, the sap wood becoming as durable as heart wood. Looking to sanitary considerations, the creosoted wooden pavement is perfect. The carbolic acid contained in the oil is a powerful disinfectant, and as the pavement described will not absorb any deleterious substance from the surface, it has only to be kept clean to maintain the best sanitary condition. This is far from being the case with wooden pavements laid on the American plan. They soon become a mass of decaying vegetable matter, and, as their powers of absorption increases with their disintegration, they become filled with corruptible matter absorbed from the filth of the street, and as their surface becomes filled with holes, it is absolutely impossible to keep them properly clean.

A good wooden pavement is also an inexpensive one. The cost, including a cement concrete foundation, 6 inches deep, would not exceed $3.00 per square yard. The system of maintenance adopted in London, of making it a part of the contract of construction, would insure good workmanship in laying the pavement, and a good permanent roadway afterwards. It would not be difficult to find responsible and honest contractors willing to take such a contract at a fair price.

In considering this subject, one should not overlook the statistics of accidents gathered in London by Col. Haywood,* which show that a London horse will travel on granite 132 miles, on asphalte 191, and on wood 446 miles, before an accident occurs.

The actual wear of wooden blocks is very slight, as long as the fibres of the wood are sound. Mr. North states that it is ⅛ of an inch per

* See full reports in the library of the Society.

annum in the streets in London, with the heaviest traffic. Mr. Geo. Frederick Deacon, Member Inst. C. E., in a paper read before the Inst. of C. E., states that, in Great Howard street, Liverpool, which is a shop street, with a traffic consisting chiefly of carriages, amounting to about 94,000 tons per annum per yard in width, the pavement was worn to the extent of $\frac{5}{8}$ of an inch in four years. This would give a life of nearly twenty years before the blocks would be reduced from 6 inches to a thickness of 3 inches, which is still sufficient to maintain the blocks in place.

In Oxford street, in London, where the traffic is equal to 300 tons per foot per day, the amount of wear has been found to be from $\frac{1}{16}$ to $\frac{1}{8}$ inch during three and a half years. This street is laid with the Henson pavement. This slight wear is largely due to the fact that the ends of the fibres do not broom, and thus retain their original strength.

C. SHALER SMITH.—I merely wish to ask Mr. Andrews—speaking of the foundations of wooden pavements—if he is aware of any pavement being laid as upon the Cleveland Viaduct, that is, Nicholson pavement upon an iron foundation ?

E. R. ANDREWS.—I am not aware of any except the Broadway Bridge in South Boston, where it was necessary to have a light pavement. A bituminous concrete about two inches thick was spread on the top sheeting and allowed to become solid ; then a thin coating of hot tar spread evenly, and creosoted spruce blocks, injected at my works with 12 lbs. of oil per cubic foot laid in rows $\frac{1}{2}$ inch apart, and the interstices filled with pitch and the surface spread with gravel.

G. BOUSCAREN.—Can you give the cost of creosoting ?

E. R. ANDREWS.—$12.00 to $16.00 per thousand feet, board measure.

G. BOUSCAREN.—Can spruce be treated well ?

E. R. ANDREWS.—Spruce does not absorb oil readily on account of the compact character of its fibres, yet it will take in a gallon of oil per cubic foot ; hemlock, pine, both white and yellow, and porous oak, are more absorbent. Wood which is the most destructible, because it absorbs water readily, is really the best for creosoting, as, for instance, the gums and cottonwood.

G. BOUSCAREN.—Have you any special rule for determining the amount of carbolic acid in the oil ?

E. R. ANDREWS.—I have not taken any pains to ascertain. The quantity depends upon the character of the coal from which the gas was made, varying from 5 to 10 per cent. It has been ascertained, however through careful experiments by a Belgian chemist, that the wood-preserving qualities of creosote oil are due rather to the water-proofing imparted to the wood by the hydro-carbons contained in it than by the carbolic acid. The latter is very volatile, and were it not retained by the gummy, resinous oil, would quickly escape into the air. In England no reference is made to the quantity of carbolic acid contained in dead-oil to be used in the specifications for contract work. Carefully conducted experiments of my own with pieces of yellow pine, 8 inches by 8 inches and 9 feet long, have shown that six months after treatment they did not absorb any water during a soaking of 48 hours under water.

M. MERIWETHER.—We made in Memphis, in 1867, what has proved to us a very costly experiment in wooden pavements. We laid there in that year and the succeeding spring some 225,000 square yards of what is known as Nicholson pavement. The streets on which this was laid rise six inches from the side to the centre of the street. Two inches to two and a half of sand were laid down, and that covered over with an inch board longitudinally. Those boards were immersed in hot coal-tar. After that were laid cypress blocks eight to twelve inches long, with a thickness of three inches, with a cleat between them, tacked to them. About one-half of that kind were immersed in boiling tar. According to the contract it was to be red cypress wood, which is very durable aboveground, and of more than ordinary durability underground, and is found to last better than ordinary oak; but the sap portion of that wood is very perishable, rotting within one or two years if exposed above-ground. The heart will last twenty-five years, or so, in fences. I mention this to show the durability of the material. Most of that was entirely covered with boiling coal-tar and small gravel, and the interstices filled with sand, making a beautiful pavement for about three or four years. Some portions of it, however, began to show decay in three years, where the sap of the block was put in. The contractor got in a good many sap blocks, and as soon as they began to decay the adjacent blocks were loosened, so at the end of about five years it was in many places impassable, and some blocks were entirely destroyed. The remnants of that wooden pavement of 1867 and 1868 are now there. The pavement cost us $3.89 per square yard. I am satisfied had the blocks all been heart cypress it would have been good for perhaps ten years. We have portions of the pavement now where the heart of the cypress was used, where there is no surface wear perceivable. However, we are satisfied with that experiment, and I do not think we will ever put down another square yard of wooden pavement of any sort. We are now making arrangements to pave with granite.

E. R. ANDREWS.—It seems to me very clear that the reason why the cypress pavement did not last was because the blocks were dipped in tar. It is not at all probable that these blocks were perfectly seasoned, because seasoned lumber cannot be found in this country for paving purposes, hence the sap enclosed within the wood by the tar soon fermented and the fibres rapidly decayed. If the blocks had been laid without being dipped in tar I think you would have had a very fair pavement now.

I would like to ask as to the condition of a pavement in Washington called the Flannigan pavement, with cypress blocks sawed from round sticks and laid promiscuously, large and small together, the spaces being filled with pitch. I have understood that that pavement has stood well.

J. E. HILGARD.—That pavement has done the best of any in the city. It was tried as an experiment on Third street, near the railway.

The city of Washington has made very extensive experiments in the matter of wooden pavements. Nothing was done to preserve the blocks. They were hemlock; none of it has lain in tolerable condition over four years; much of it had become intolerable even before that time. I think the work was badly done from there not being proper supervision. An immense amount of work was undertaken to be done within a limited time. It was the most disgraceful failure of wooden pavements ever known, and it has been a case of unprecedented decay. None of the streets were in a fit condition to travel over after four years; most of them have been replaced by concrete pavements. With us the experiment has been a very expensive one. Climate may have had something to do with it, for it is very warm in summer and we have very frequent showers, but certainly the decay was unusually rapid.

E. R. ANDREWS.—In 1869, Columbus avenue, in Boston, was paved with wood; every one who had any patent pavement was allowed to put down a piece; one section was laid with creosoted blocks, but very imperfectly prepared. All the pavements were taken up and the street repaved with the Trinidad bitumen in 1877; but a small delta of the creosoted pavement was left, which is still sound and in good condition.

M. MERIWETHER.—I did not mention that about one-half of the pavement of which I spoke was laid with planks dipped in boiling tar, and after making about one-half of it the very difficulty suggested by Mr. Andrews arose, and we ceased to immerse the blocks, upon the theory that if the under side was covered with tar it would cost more, so they stopped that process and laid the balance in the other way, and after the planks were down they covered the surface simply with coal tar, upon the theory that the bottom of the planks, not being concealed, the acid would go down; but we did not discover that it made any perceptible difference. It would probably have been satisfactory if the blocks had been heart wood, but the sap of the blocks decaying, led to their destruction by the wheels passing over it. It was an utter impossibility to have seasoned wood for such an extent of pavement—some 225,000 square yards, laid at once. The wood was not in the market, and no one could afford to keep such a stock on hand. The result was that the wood was brought directly from the mills and put down within two or three months from the time it was taken from the stump. I do not think any process short of thorough seasoning, or some process of drying quickly by steam, would do any good. We found it impossible to get the heart cypress entirely, and the sap wood would decay in two or three years.

The blocks covered the surface and prevented the water passing down, but it may have passed on the side. They merely covered the surface with tar, and some little water might find its way down the side of the block. The spaces between the blocks were thoroughly rammed with small gravel, with sand with it, and the surface covered with coal tar.

C. SHALER SMITH.—I have recently made some experiments for the St. Louis Bridge, which illustrate in a marked degree the action of those preservatives depending on carbolic acid for their antiseptic value. Finding the wooden stringers of this structure beginning to rot at the ends and other points of support, while the remainder of each stick continued sound and untouched by decay, I tried to arrest the rotting by the injection of creosote containing ten per cent. of carbolic acid into all these timbers which showed signs of decomposing fibre. The effect was remarkable. Sound wood was unharmed, but where decay had already commenced the acid seized upon the wood and converted all parts affected by rot into a brown cinder, in many cases absolutely destroying the bearing value of the stick. The experiment was extensively and exhaustively tried, and I am satisfied that while creosote is excellent when properly applied to perfectly sound lumber, it will not arrest decay when once started, and in many cases will destroy all the unsound parts of a stick.

And also, that in the use of cresote, the proper proportion of carbolic acid is a very important element, and should be fixed by specification whenever this system of treatment is used. I have likewise continued experimenting on other methods of preserving wood, three of which have given good results.

The first is the " Thilmany, old process." This consists in impregnating the wood with sulphate of copper, and subsequently with the chloride of barium. The chemical action of the two salts fills the pores of the wood with the preservative chloride of copper, mechanically fixed in position by the insoluble salt, sulphate of baryta.

The second is the " Thilmany, new process."* Here the first impregnation is sulphate of zinc, the second chloride of barium, and the resulting salts, chloride of zinc fixed as before by sulphate of baryta.

The third is known as the tan-zinc process. The first impregnation is with chloride of zinc dissolved in a solution containing $2\frac{1}{2}$ per cent. of glue. This is followed by an injection of a tannin solution which precipitates the glue, forming tannate of gelatine, a perfectly insoluble com-

---

* For specifications of this process, see Appendix No. 6.

pound, and which fixes the chloride of zinc so thoroughly that it cannot be extracted either by boiling or steaming. I am not now prepared to state which of the three processes is the best, as our experiments are still going on. It may be safely asserted, however, that no system of treatment depending on a soluble salt, as in the Burnett or Boucherie processes, is of any value unless the salt is fixed in the wood by a subsequent injection which will fill the pores with an insoluble substance. I have procured specimens from various Burnettized bridges, and the analysis has shown in every case that the zinc had been entirely washed out of the wood. In treating wood by either the cresote or metallic salt systems the antiseptic injection is virtually worthless unless the wood has been previously deprived of its sap. In doing this the following rules should be rigidly observed :

First.—The steam bath should not exceed 5 pounds pressure, or 240 deg. F. in temperature, and the lumber should remain in the bath for not less than 90 minutes for sticks under ten feet in length, and 9 minutes additional for every additional foot of length.

Next.—The steam bath should be followed by an exposure to a vacuum of not less than 11 pounds pressure for 40 minutes for sticks less than 10 feet in length and 6 minutes more for each additional foot, after the vacuum is reached.

Last.—The preservative injection should be run in while the vacuum is still on, and after the cylinder is filled the injection pressure should be brought up gradually to not less than 100 pounds. The time for its continuance will vary with the wood used and the length of the stick.

By the observance of these rules in treatment and the selection of thoroughly sound lumber, it is in my opinion perfectly practicable to produce a good wooden paving block, which, when properly laid, will make a clean and lasting pavement. Treated gum blocks placed in the testing machine at the St. Louis Water-Works stood the passage of 95,000 wheels with a wear of only one-eighth of an inch. The wheels of testing machine were loaded to 2,000 pounds per wheel, or 800 pounds per inch of tire width.

It is hardly necessary to reiterate, however, that no system of treatment, however good, will arrest decay, convert unsound blocks into sound ones, or render blocks cut from dead trees fit material for a pavement.

E. R. ANDREWS.—I do not wish to advocate the use of decayed timber, or to intimate that by the process of creosoting, wood which has lost its quality by reason of decay can be restored to its original strength ; but if partially decayed timber be creosoted the process of decay is arrested, and there may be cases where it will be advisable to creosote it, and thus save for future usefulness timber which would otherwise be valueless.

As an instance of such an experiment, I quote from the *Journal of the Society of Arts*, London, June 1st, 1860, containing a paper by G. R. Burnell, entitled "On Building Woods, the Causes of their Decay, and the Means of Preventing it," and a discussion thereon by the members of the society, during which Mr. John Bethell made the following statement (see page 565).

"That in timber where decay had commenced it had been stopped by the injection of creosote. He could confirm that fact by stating that about twelve years many thousand sleepers were packed upon the Lancashire & Yorkshire Railway some time before being used. When they were about to be used they were found to be more or less decayed, and it was a question whether the whole should not be sold for firewood, when it was determined to submit the sleepers to the process of creosoting. After those sleepers had been down for ten years it was found that not only had the decay been arrested, but the sleepers were as good as if they had been sound, new timber."

I do not quite approve of General Smith's specifications for the treatment of timber. Unless he has positive proof to the contrary, I should doubt whether large-sized timber can be properly dessicated in ninety minutes. In my own experience I find that the time required is greater with large timber than with small, and in proportion to its square rather than its length. Wood is a slow conductor of heat, and a 12-inch by 12-inch stick cannot be heated through to a point of vaporization in ninety minutes, and the moisture cannot be withdrawn until it is vaporized. Moreover, it will require a vacuum of 20 to 25 inches during several hours to withdraw the vapor, so that danger of decay from moisture within the wood shall be removed. If this is not accomplished before injection with creosote, the effect will be to close up within the wood fermentable substances. Thorough injection cannot be effected. The incompressible water will not permit the oil to permeate those portions of the wood where it exists.

Engineers should give time enough to do good work. In the practice abroad and in this country, where creosoting is done intelligently, the work is never hurried. Large piles and square timber cannot be properly dessicated and creosoted in less than from 20 to 24 hours. Such work costs more money, but is cheapest in the end. Paving blocks can be treated much more rapidly. During the infancy of creosoting in this country great care should be taken to secure thorough work in order to obtain for such timber the longevity attained in Europe and the confidence of consumers.

DAVID E. McCOMB.—I shall confine my remarks to granite and bituminous pavements as laid in the cities of Washington and Georgetown, D. C.

The standard stone pavement has a base of six inches of hydraulic concrete, upon which is spread three inches of sand, in which the granite blocks, measuring eight inches by four inches by six inches in depth, average, are bedded with close joints, which joints are filled with screened gravel, pea size, heated to a temperature of 400 deg. F. The blocks are then brought to a solid bearing by the use of the ordinary rammers, after which the joints are filled with coal tar refined to 400 deg..F., which removes the light oils, water, etc., yet retains the heavy oils, "cut back," which is the result of a mixture of the residuum of the destructive distillation of coal tar, with still bottoms, being carefully guarded against. After the joints have been filled with the tar, as before described, fine heated sand, or perfectly pulverized limestone, is spread over the surface, which completes the pavement. In situations where the sub-foundation is solid and unyielding, the concrete base is dispensed with, six inches of gravel, compressed by a heavy steam roller, being substituted in its stead.

All the different kinds of tar pavements have been laid and tested in Washington, and our experience is that they are not economical, requiring too extensive repairs and too frequent renewals.

The only two pavements that have given any reasonable degree of satisfaction are those having a mixture of Trinidad bitumen and coal tar, refined to 400 deg. F., in approximately equal proportions as the cementing medium of the sand, limestone, or other ingredients forming the body of the wearing surface. Such a top coating is good for about seven years, requiring, however, watching and small repairs during this time, after which it seems impossible to patch it successfully.

There have been laid six squares of Neuchatel under two different contracts, the first in 1872, the latter in 1876. That laid in 1872 on I street has stood the test of time and limited travel very well, and is in fair condition now. That laid in 1876, on Pennsylvania avenue, is in very poor condition, and requires extensive repairs. This class of pavement possesses one fatal objection, viz., its extreme hardness, as a consequence of which, when the surface is covered with a film of water, it is only by the exercise of great care that horses can travel upon it without slipping and falling oftentimes, especially in turning corners.

The pavement of this general class that has given the most satisfaction has for its foundation a depth of six to eight inches of hydraulic concrete, upon which is spread a coat of asphalt mastic half inch thick, which is intended to give an uniform surface to compress the top coat upon. This cushion coat, as it is termed, is composed of 62 parts of fine sand, 15½ parts of pulverized carbonate of lime, and 22½ parts of asphaltic cement; all the compression given to this coat is that due to rolling with a hand roller weighing about ten pounds per inch run. The top coat, or wearing surface, is composed of 65 parts of fine sand, 16 parts of pulverized limestone and 19 parts of asphaltic cement. The sand is required to contain not more than five per cent. of clay, and to be a little finer than would be usually accepted for use in making mortar ; the asphaltic cement is composed of refined Trinidad asphalte and refined petroleum still bottoms, or paraffine oil in the proportion of 100 pounds of the former to 19 pounds of the latter. The pulverized carbonate of lime is mixed, when cold, with sand heated to 300 deg. F., which mixture is then mixed with the asphaltic cement, heated also to 300 deg. F., and a thorough incorporation is effected in a twin pug mill, after which the material is carried in carts, having an arrangement to keep it hot, to the work, and is spread upon the cushion coat with rakes, and having a thickness of 2 8-10 inches. It is then rolled with the hand rollers before referred to, the jointings at the curb being tamped with the pilon described by Mr. North, which is heated to a temperature of 500–250 F., the test of its being too hot is that it scorches a white pine plank, on hand for the purpose of trial. The surface is then rolled with a steam roller weighing 300 lbs. per inch run, of large disk. No difference is observable between rolling with cold or heated roller. The disks of the roller are kept moistened with crude petroleum, which prevents any tendency to pick up the top coat, and when the proper proportions are observed no trouble is experienced from the machine shoving the material ahead, this occurring when the mixture is too rich in asphaltic cement. After being compressed, the thickness of wearing surface is two inches.

The gutters are paintwithed asphaltic cement ironed in, the object being to prevent the degradation that occurs at this portion of the carriageway, caused mainly by the fact that this portion of the carriageway does not share with the traffic compression produced at the other parts of the roadway.

Trouble has been experienced on pavements alongside of street railroad tracks, the cross ties of which, if not well ballasted, move up and down sufficiently to break the pavement over them and leave ugly ridges. The tendency to cut into ruts alongside of the rail is counteracted by laying granite blocks, eight inches by four inches by six inches, alternately as header and stretcher, bedded in mastic, the toothing thus formed obviating this tendency to rut.

The above described pavement, when honestly proportioned and laid, possesses all the elements which go toward making a good pavement, and it is not expensive, costing at the present time about $1.75 per square yard.

I agree with Mr. North that a bond for maintenance for a term of years should be required from the contractor for this class of work, there being so much of honesty and skill required in refining and manipulating the materials composing it.

E. P. NORTH.—I would like to ask Mr. McComb how much clay is left in Trinidad bitumen and how it is refined.

D. E. McCOMB.—There is supposed to be left in the refined bitumen twelve per cent. of impurities, it being practically impossible to refine it so that it will have a smaller percentage of impurities than that.

NOTE ON THE NOMENCLATURE OF BITUMENS.
BY EDWARD P. NORTH.

While such strenuous efforts are being made for uniformity in the matter of measures, the nomenclature of bitumens should secure attention.

In Paris, and in France generally, the nomenclature of M. Malo and others, as given in Transactions, Vol. VIII., page 121 (May, 1879), is used. Colonel Haywood also employs it in the specifications and reports of the City of London, and it is believed that all dealers and manufacturers of asphaltes use the same nomenclature.

On the other hand, in this country dealers in Trinidad bitumen and its mixture almost invariably call their compounds asphalt or asphaltum. Tar pavers and roofers also apply the same names to their products, both calling such asphaltes as are capable of compression Neuchatel, apparently because the Val de Travers asphalte happened to have been imported into this country by a branch of the "Neuchatel Asphalte Company, Limited," of London, which bought out "La Societe Generale de Asphaltes de Suisse," but could not take the name Val de Travers, as there was at that time an organized company in London bearing that name. Val de Travers asphalte is not called Neuchatel in any part of Europe, besides which there are at least three other asphaltes that can be compressed, besides many others that are used for mastics.

As the distinction between asphalt or asphaltum on the one hand, and asphalte on the other is too slight to attract attention of any but a critical reader, the advisability of following Malo's nomenclature is submitted to the Society as the most logical and convenient.

F. RINECKER, of Wurzburg, Germany (through the Secretary).—Besides the methods of paving described in the paper of Mr. North, the following may be worthy of a short mention :

Brick pavement, especially in Holland, where very hard brick, *Klinker*, is used for this purpose.

Flag pavement, in many Italian towns, consisting of large slabs, laid in rows for the wheels to run over, the balance of the street being common pavement.

Concrete pavement, which (if I am correct) was tried in Paris and New York. I am ignorant of the results, however. I have seen some sidewalks and depot platforms in Germany, which wear well. For traffic it will be worthy of consideration, whether the broken stone or gravel used in the concrete should not be of the same hardness as the cement to insure equal wear. In using harder stone the unequal wear might be a cause of failure.

In selecting any system for covering roads and streets uniformity is necessary throughout a whole city, at least in its main thoroughfares. The safety of horses greatly depends on their shoes, but the patterns of the shoes differ with the roads on which they are to be used. The connection between shoes and roads needs no comment, and in my opinion it is a grave mistake to use stone in one street and wood in another, when the same traffic has to run over them both.

I attach some notes taken from German and French periodicals, the figures of which I have reduced to United States standards.

As to the cost of some pavements to the square yard, the *Deutsche Bauzeitung*, 1877, has the following :

	Vienna. Granite.	London. Granite.	Asphalte.	Wood.
Durability, years	85	15.6	17	11.34
Cost of construction	$3.07 to $3.37	$3.77	$3.80	$3.20
Total cost of maintenance	1.74	1.40	3.27	3.37
Aggregate cost averaging to one year.	18	.37	.34	.62

From experience at Buda Pesth the following data for 15 years are furnished :

	Granite on Broken Stone.	Concrete.	Trachyte on Broken Stone.	Wood.	Asphalte comprimé* on Concrete.
Thickness of pavement, inches	7.1	7.1	7.1	?	2.36 to 2.76
" " foundation	6.3	6.3	6.3	?	9.7
Cost of construction	$4.50	$5.05	$5.50	$3.00	$4.35
Total cost of maintenance	2.40	1.80	3.00	6.00	3.00
Aggregate cost of 15 years, averaging to one year, cents	.46	.45	.57	.60	.49

The paving stones of Paris are described in " Romberg's Zeitschrift f. pract. Baukunst," 1878, as follows :

According to the specifications a distinction is made between large and small blocks.

Large blocks to be 7.9 to 9.2 inches long, 6.3 to 9.2 inches wide and 7.9 to 9.2 inches high.

Small blocks to be 6.3 inches long, 3.9 inches wide and 6.3 inches high. The latter, *"paves de petit echantillon,"* are preferred of late. Differences in size to 0.4 inch are admitted.

On account of the bond, a certain percentage to be blocks one and one-half times as long as specified above, *"boutisses."*

According to the dressing, two qualities are distinguished : smooth and rectangular blocks for joints of only 0.2 inch, and rough ones for joints up to 0.6 inch.

The material is sandstone, *"gres d'yvette des Vosges,"* or *"de l'Ourthe,"* and porphyry from Belgium and Bavaria. This porphyry, however, has not given satisfaction, wearing too smoothly.

The price is $44 to $125 the thousand, and $8 to $13 additional for dressing.

The Macadam is being replaced where annual repairs exceed 50 cents the square yard.

* Asphalte comprimé is the French term for the compressed powder.

As to the cost of the Paris pavements, the following notes are from "Annales des ponts et chaussées," 1877 and 1878 :

	Pavement.	Macadam.	Asphalte.
Cost of construction	$2.55 to $8.45	?	$1.93 to $2.42
Annual cost of maintenance, cents	9.6	29.0	17.7

The wear of Macadam, amounting to 23.37 cubic yards to the mile and 100 horses.

Annual cost of maintenance to the square yard :

	1872–5.	1876.	1878·
Pavements, cents	7.7	8.2	8.5
Asphalte, "	19.3	20.9	20.4
Macadam, "	29.0	34.0	82.2

The watering with carts costs :

On Macadam	19.3 cents to 1,000 square yards.
On pavements	9.6 " " "

J. B. Dumas, Assistant Engineer of the City of Paris, published comparative estimates in " Nouv. Ann. de la Constr.," 1878–79, from which the following data are compiled :

	Price to the yard.
Cost of construction.	
Pavement of rectangular blocks " *gres*," 8.9 by 6.3 by 6.3 inches	$3.22
Macadam of silex	1.61
" " meuliere	1.84
" " porphyre	1.98
Asphalte comprimé, 2 inches thick on 8.9 inches concrete.	
Using for the latter hydraulic lime	2.93
" " Roman cement	8.08
" " Portland	8.15
Each inch of asphalte above 2 inches additional	1.05
Wood pavement, Trenaunay	2.76
" Norris	4.31
Annual cost of maintenance and repairs:	
Rapaving	24 to 64 cents.
Asphalte, roadway	21 "
" crossings	82 "
" sidewalk	4.8 "
Macadam, silex	47 "
" meuliere	55 "
" porphyre	129 "
Wood pavement	56 "

Both systems of wood pavements failed in streets with high traffic.
Experiments with asphalte coulé* have shown its unfitness for wagon traffic.

THE ENGINEER, 1878, Vol. XLVI., p. 358, ascribes the invention of the Macadam to John Lochhead in 1794.

J. J. R. CROES.—Referring to the amount of water used for keeping down the dust on Macadamized roads (p. 82), the following may be of interest.

In the year 1872 an account was kept of the water used on the drives in the Central Park in New York City.

The length of carriage ways is 9.435 miles, of widths from 30 to 60 feet, averaging 54 feet. The area occupied, including spaces for carriages in waiting, is very nearly 250,000 square yards. The carts used for watering are in the form of a segment of a cylinder of 35 inches diameter,

---

* Asphalte coulé (poured) is the French term for mastic.

the height being 26½ inches and the chord 20 inches. The barrel is 90½ inches long, and contains 40.7 cubic feet of water.

During the season of 1872, from April 1st to October 31st, the roads were sprinkled on 136 days, using 81,305 barrels of water, or 3,309,114 cubic feet, an average of 24,332 cubic feet per day, or 97⅓ cubic feet per day for each 1,000 square yards. The greatest amount used in any one day was on July 1st, when the temperature ranged from 77 degrees to 93 degrees F., and 929 barrels or 37,810 cubic feet of water were used; an average of 151¼ cubic feet per 1,000 square yards.

The next greatest amount used was on June 22d, when the temperature ranged from 70 degrees to 86 degrees F., and 890 barrels or 36,223 cubic feet of water were distributed; an average of about 145 cubic feet per 1,000 square yards.

At least one-half of the area watered was sprinkled twice as often as the other half, in consequence of its greater exposure and the greater travel upon it.

I am informed that during the summer of 1879 the carriage way of Fifth avenue from Twenty-third to Thirty-fourth streets, 0.55 mile in length and 40 feet wide, embracing 12,907 square yards, was kept watered by six carts holding 70 cubic feet each, and making from three to six trips per day. This would make the amount of water used from 97.5 to 195 cubic feet per 1,000 square yards. The pavement is of trap blocks.

E. B. VAN WINKLE.—Referring to Mr. North's paper: The form of roller proposed for the upper layers of earth roads, namely, large and smaller sized disks placed alternately on the axis of the roller, I have seen used with excellent effect on an embankment, and it could doubtless be so used for the foundation of earth roads, but is unsuitable for surfacing and road maintenance. The ridges left by the use of this form of roller would tend to carry surface water longitudinally instead of the shortest distance transversely to the gutters.

I believe it would be found very efficacious and economical in the maintenance of earth roads to have the regular passage over them—one or more times, according to their importance—of a moderately heavy two-horse roller a few hours after the cessation of rains. About one roller to every twenty miles of ordinary road should be kept provided.

Referring to page 65: I should think it questionable whether the addition of hard pan to a clay road would be of use.

Referring to pages 68 and 69: It would be interesting to know the results of a combination of the first and last of the systems of rolling Macadam roads enumerated by Mr. North, that is, steam rolled and traffic made. I would suggest that the steam roller be first used without binding material, or with very little, to bring the road metal to a passable surface, and then open the road to traffic.

My experience coincides with that of Mr. North—that the hardest kind of stone, if of the proper size, will produce a firmer roadbed when traffic made, without softer binding material.

Referring to page 71 : The reason why the Macadam in Basnat street, Liverpool, does not show the wearing qualities of a well-puddled trap road I conceive to be not on account of the binding material—coal tar, pitch, &c.—but owing to imperfect consolidation due to hand rolling. I have no doubt that, the rolling being equal in all cases, a binding material of pitch as described would give better results than where cement, clay, or "hoggin" were used, as less water would reach the foundation, and the pitch would have more elasticity, giving somewhat without breaking or crushing.

Referring to page 72 : M. Malo's picturesque description of the Macadamized streets of Paris will apply with equal force to London or New York.

As far as I have any experience, Macadam pavements are a failure for city streets, except in some isolated cases where the traffic is merely nominal, or where all other considerations except pleasure driving are out of the question.

Referring to page 78 : One great reason for the unexpected success of Mott avenue was, in all probability, that the heavy rolling over the saturated fresh filling of earth of the road (some four or five feet in depth) most thoroughly compacted it, hence giving subsequently a very solid foundation for the Macadam.

Referring to pages 81 and 82 : The transverse section of a pavement is most important. The general tendency is to give too much crown to the pavement. In streets where there are horse railroads this crowning has generally become excessive, the trackmen tending always to raise the track regardless of any established street grade, while the city employés who repair the pavements meekly assume that the track is at grade and pave flush with it, while at the same time the curb is maintained at grade or settles below it. It would seem proper that the authorities in charge of street roadways should, at the time a railroad is being built, determine a proper profile for it. This profile should be filed, and the railroad forced to conform to it.

Another fallacy is, that an excessive amount of crowning adds to the strength of the pavement on the principle of an arch, the paving blocks being the voussoirs and the curbstones the skewbacks. The absurdity of this illusion is readily apparent if we conceive of an arch of 30 to 60 feet span with a versed sine of one foot or less, skewbacks half a foot thick resting on compressible earth, and, more wonderful still, the voussoirs in contact only at the extrados. The greater the crown given to a pavement the less will the depressions resulting from poor paving be noticeable on account of holding water.

In an earth road the transverse slope should never be less than the longitudinal slope. The restriction I consider unimportant in a stone block pavement, as with this class of pavements there is no danger of ruts being formed by running water, while it is quite desirable to have the rainfall scour quite an extent of the pavement surface before reaching the gutters.

Gillespie says that the proper section of roadway surface should be formed by two planes inclined from the gutters upward towards the centre of the roadway, with their intersection rounded by a slight curve.

That this is correct I have tried to believe for a long time, but observation convinces me that it is not the best form. The arc of a circle is practically the best cross section for street pavements. If not from choice, at least from necessity, the bulk of travel is along the centre of the street, the portions on each side next the curb being occupied by vehicles standing. With a section of the arc of a circle the centre of the roadway is almost level. Another advantage is, that while the gutters are running full, the width of deep water is narrower than when the surface is an inclined plane.

Referring to page 83 : I fully agree with Mr. North as to the excellence of granite pavements in England. London, particularly, excels in the quality of its stone block pavements. I have seen nothing elsewhere equal to them for streets of heavy traffic. Their great superiority arises principally from their concrete foundations. It is surprising that similar foundations are not always used when the streets in the business portion of the city are relaid with stone blocks. The filling of the joints between the blocks with pitch I think would be a decided improvement upon the London plan of grouting.

Referring to page 84 : Wood pavements, as far as I have any experience, have been without exception a failure. I can conceive that creosoting the blocks and filling the joints with bitumen would prevent decay, and thereby materially lengthen the life of this class of pavements, but, so far, nothing has been—and in all probability never will be—invented that will prevent wood, when subjected to an incessant impact and attrition of iron horseshoes and heavily loaded steel tires from acting as wood—that is, quickly wearing down. The pits in the surface of wooden pavements have been proved by borings to be due much more to the actual wearing off of the wood than to settlement.

Referring to page 97 : My recollection of those sidewalks in Paris covered with asphaltic mastic ("bitumen" so called) is, that they are not very satisfactory. During July and August almost every footstep left an impression in the mastic, and in winter they were excessively muddy, and had generally sunk out of plane.

Have any experiments ever been made with the bituminous limestones·

found in this country, to use them as paving material in the same manner as the regular asphalte ?

JOHN BOGART.—Experiments were made some years ago in Chicago with a peculiar limestone, which it was thought might be utilized in a manner similar to the imported asphalte. The result was a good Macadam pavement, but apparently the impregnating matter had no advantageous effect—my recollection is that it was petroleum rather than bitumen, and that it did not exert any cohesive force.

C. C. MARTIN.—The Smith Hydraulic Stone Crusher as now manufactured possesses *two* peculiarities which render it superior in certain respects to former patents :

First.—The power is applied through an hydraulic cylinder, which has eonnected with it a safety-valve.

Second.—The opening through which the stones pass after being crushed is wider at the bottom than at the top, as shown by the cross-section at $c\ c$.

These are the two essentially novel and good features of the machine; all of the rest are simply mechanical arrangements for transmitting the power from the engine to the machine, and may be arranged as here shown or in any other way. In this machine a is the hydraulic cylinder, b is the frame, c is the movable jaw, d the toggle, e the hydraulic ram, f the plunger, g the connecting rod, h the band-wheel, k the fly-wheel, l the stationary jaw ; the section $c\ c$ is a front elevation of the stationary jaw. When the machine is to be made ready for use the ram, e, is drawn back so as to place the movable jaw, c, far enough from the fixed jaw, l, to permit stones of the largest required size to pass through between them ; the plunger, f, is drawn up out of the hydraulic cylinder, a, and the cylinder is then filled with water or other liquid, and the safety-valve is weighted. The operation of the machine is as follows : Stones are thrown between the jaws and the engine started ; as the crank shaft revolves the plunger, f, is forced down into the liquid, and, displacing a portion of it, forces the ram forward, which, through the toggle, d, presses the movable jaw against the stone, which is crushed. The upward movement of the crank withdraws the plunger, and the ram is drawn back to its original position, and the jaw is again opened : thus, every revolution of the crank shaft produces one stroke of the jaw. The machine works well, running at two hundred to two hundred and fifty revolutions per minute, and will readily and regularly break seventy-five cubic yards of limestone in ten hours to sizes varying from $1\frac{1}{2}$ to $2\frac{1}{2}$ inches largest dimensions.

The advantage of the safety-valve is, that in case any stone which cannot be broken, a sledge-hammer dropped from the hands of a careless workman, or any other unyielding body, gets between the jaws, the only effect is to force open the safety-valve and permit the escape of a portion

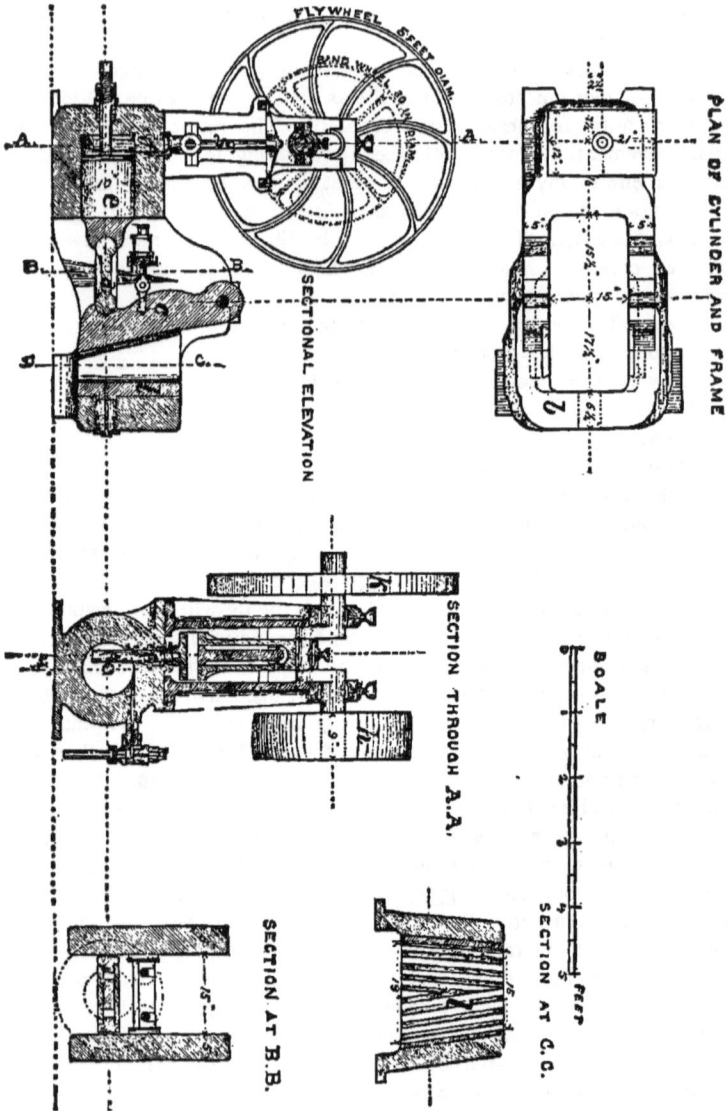

HYDRAULIC STONE CUTTER (SMITH'S PATENT).

discharged is received in a small tank, and the continued motion of the machine pumps it back into the cylinder, and the crushing goes on. If of the liquid from the cylinder, thus stopping the motion of the jaw without stopping or changing the motion of the fly-wheel, The liquid thus the obstruction still continues, the safety-valve is again opened. The result of this safety attachment is that the machine is never broken.

The advantages of widening the jaw at the bottom are twofold, as will appear from a consideration of the operation of the stone in the crusher. The broken stones occupy more space than the unbroken, and the smaller the pieces the more space do they occupy. The stone entering at the top of the jaw is broken, and each successive stroke crushes it more, and in the old style of jaw the stone became clogged and many pieces were ground to powder before they could escape, thus wearing out the jaws and cheek pieces, and consuming unnecessarily a large amount of power. The widened jaws permit the stones to escape as soon as they are broken, thus leaving them of more uniform size, making less dust, and breaking them with the least possible expenditure of power.

EDWARD P. NORTH.—Regarding objections made to certain clauses in the paper under discussion, it may be said that the roller with unequal sized disks was not advised for maintenance, but unless it was so guided that the larger disks always followed the same tracks its use would cause less tendency to longitudinal ruts in the road than the ordinary traffic.

The advisability of surfacing a clay road with hard pan must depend greatly on the character of the hard pan—a clay hard pan might be objectionable.

Messrs. Spielman & Brush's practice, as detailed on p. 119, is undoubtedly sound under the circumstances, viz.: an absence of water for compacting and puddling, but a better road, with less tendency to internal wear, would have been formed by a steam roller, if sufficient water could have been procured.

Lavoinne seems to have misunderstood the effect of the screenings, *i. e.*, small fragments of stone from the breakers. The coarser of these are applied only when compacting has proceeded so far that the stones have a tendency to roll over one another and round their angles; these screenings bind the stone and prevent to a great extent further wear. About 33 per cent. of screenings are worked into the metal and are used in puddling. Probably about 25 per cent. of this is coarse screenings in the interstices between the stones, the rest being fine screenings and that used in puddling.

Law & Clarke, p. 145, give 10 or 11 cubic feet interspace per yard—40 per cent. in compacted stone. In Paris, about 24 per cent. of sand is used in binding, and, as above stated, 33 per cent. of screenings are used here.

It was not the intention to argue for poorly constructed wooden pavements, but rather to show the practice where wood pavements are successful, with the hope that it might be followed in this country. The fact that wood pavements as usually constructed here, and never repaired, have proved failures, cannot be denied, but that a well made and intelligently maintained wood pavement would be a failure is doubted.

The following memorandum from Cleveland confirms this view : " In the year 1873 5,070 square yards of creosoted wooden pavement were laid in Franklin street, Cleveland. The pine blocks were creosoted with about 4 pounds of oil per cubic foot. Since that time no money has been expended in maintenance, and the pavement is in excellent order.

"At the same time, and in the same street, an equal area was laid with blocks prepared by the ' Thilmany' process ; of this, 1,400 square yards have been relaid by the city."

It is unfortunate that the asphalte pavements in Washington were so poorly laid.

I street has such an excessive crown that it is only in the centre of the wheelway that the street is sufficiently flat for ease or comfort in driving.

On Pennsylvania avenue the transverse profile is good, but the heated powder was apparently laid on damp concrete, and the surface is badly cracked in consequence, and has been extensively repaired, though it was laid in the fall of 1877, and it is doubtful if it wears more than five years in all.

In Paris, with a much heavier traffic, asphalte lasts 15 years.

The bituminous mastic pavements described by Mr. McComb present very fine surfaces, and, if they wear as well as now hoped, will have all the advantages of compressed asphalte, except durability, at reduced cost, and their use may be advisable in streets of light traffic.

The present practice is to make the wearing surfaces harder than has been usual, it having been held that *all* mastic pavements, subjected to street traffic, should be made so soft as to dent in summer, to prevent their breaking up in winter.

The weak point of bituminous mastics seems to lie in the expense necessary to free the Trinidad bitumen from clay, and the difficulty of getting sand that is free from clay, and fine enough to absorb sufficient bitumen for cohesion in cold weather, without an excess in hot weather, Loam or loamy sand was used at first to enable the mixture to carry sufficient bitumen for wear, but the rapid rotting in the gutters, and at other points where water lies, has led to the use of 15 to 16 per cent. of limestone, ground to pass through a sieve with 26 meshes to the inch, which probably will not be so detrimental to the mastic as clay.

In reply to a question as to the amount of clay allowed in mastic work in Europe, M. Leon Malo writes : " In principle, no clay at all ought to be in mastic ; the asphalte rock, being quite pure limestone, is desired to be mixed with quite pure bitumen. But, in reality, the bitumen of Trinidad having been generally adopted in works by want of pure mineral bitumen, and as it is impossible to quite deprive that bitumen of its clay, it is admitted that the clay brought into the asphalte works by the Trinidad bitumen can be accepted ; that is to say, from 2 to 3 per cent. of clay in the mastic. For the compressed asphalte, to which no bitumen is added, not one quantity of clay is allowed."

Regarding the asphaltic mastic sidewalks of Paris, it should be noted that there is always a temptation on the part of the mastic workers to use too much bitumen, as the larger the percentage the more easily the mastic is worked ; and, unless it is kept very low, the pavement will dent in hot weather. The mud must have been brought on from the wheelways, as, according to Chabrier, the mastic sidewalks under the arcades of the Rue de Rivoli, Paris, wore about half an inch in 13 years—too small a wear to produce much mud.

The necessity for better pavements than those offered by the prevalent granite blocks, will justify the insertion of a translation from M. Leon Malo's " *Note sur L'Etat Actuel de L'Industrie de L'Asphalte Paris,* 1879."

The experiment of compressed asphalte for carriageways has now been made. It has shown its defects and its advantages ; we do not hesitate to say that its advantages far exceed its defects.

We do not now speak of its noiselessness nor of its precious property of creating neither dust nor mud ; of its agreeable aspect to the eye, which is a quality not to be neglected in the *ensemble* of the embellishment of a city ; we have shown the importance of these in our first work. We insist only on its action upon the public health ; a point which we did not sufficiently emphasize.

The asphalte, placed over the soil, like a layer of caoutchouc, absolutely intercepts any communication between the soil and the atmosphere ; it does not allow rain water, which runs rapidly into the sewers, to penetrate ; and thanks to its water-repelling (" *hydrofuge* ") character, the carriageways dry as soon as the rain has ceased.

A stone pavement, on the contrary, permits a constant communication between the soil and air through its joints. All the impurities of the surface, dissolved by the surface waters, are by them carried into the earth ; then, when the sun strikes it, these impure waters are evaporated, returning miasma, bred below the pavement, into the air ; it is an evil which has been long recognized and uncontested, but against which no remedy is known.

The system of joining pavements with asphaltic mastic might, perhaps, obviate it, but it is very costly, and renders the carriageway extremely hard for vehicles. Compressed asphalte seems to have solved the problem ; and we are compelled to think that of all the services which it can render, this is the most efficacious and the most precious. To be sure, the influence of deleterious miasmas radiated from the different systems of pavements with open joints, cannot be analyzed in a precise manner; it is judged rather by induction than by a direct observation which can be put in figures; but it is not the less evident and not the less to be dreaded.

M. Malo also gives the following tabulated results of analysis made in the laboratory de L'Ecole des Ponts et Chaussées on different asphaltes:

	Val de Travers.	Seyssel.	Lobsan.	Sicilian.	Maestu.	Forens.
Water and matter volatile at 212° F[1].....................	0.50	1.90	3.40	0.80	0.40	0.25
Bitumen.......................	10 10	8.00	11.90[2]	8.85	8.80	2.25
Carbonate of lime..............	87.95	89 55	69.00	87.50	9.15	97.00
Silicious sand............... ..	''	''	3.05	0.60	57.40	''
Aluminum and peroxide of iron	0.25	0.15	5.70[3]	0.90	4.35	0.15
Sulphur...........	''	''	5.00	''	''	''
Carbonate of magnesia........	0.80	0.10	0.80	0.95	8.10	0.20
Different minerals insoluble in acids............	0.45	0.10	''	''	11.35	0.05
Loss......	0.45	0.20	1.65	0.40	0.45	0.10
	100.00	100.00	100.00	100.00	100.00	100.00

NOTES.

1. The water given above depends on the dryness of the sample at the time of analysis; the figures not being of importance in the result.

2. This quantity appeared to contain a certain proportion of oil, which was mixed with the bitumen and was not exactly determined.

3. This comprises 4.45 of iron, combined with sulphur.

M. Durand-Claye, Director of the laboratory, speaking of the Lobsan, says: "It contains about 9½ per cent. of pyrites, which may become the cause of failure in employing this material. The heat to which it is submitted may cause it to lose half of its sulphur, and be transformed into protosulphate of iron, an oxydizable material, which by exposure to the air is transformed into a soluble sulphate of iron; disintegration may result from this a short time after putting it in place."

Mr. H. F. Starr, of the Columbia College School of Mines, kindly made an analysis of some Limmer asphalte, which gave the following result:

```
Bitumen.................................................. ...................   8.26
Clay........................... .... ....... ................. .............   4.98
Carbonate of lime.......................................... ................  56.54
Carbonate of magnesia........................ ....................... ......  27.01
Sesqui-oxide of iron...................... ................. . ............   3.21
                                                                            --------
                                                                            100.00
```

This is the only analysis I have seen of an asphalte that will not compress, and the only one, excepting the Maestu, in which there is over one per cent. of carbonate of magnesia.

M. Malo gives analyses of seven different cargoes of crude Trinidad bitumen. In each case the samples were thoroughly dried, losing from 32 to 38 per cent., after which the average of the analysis was bitumen 51½, and clay 48½ per cent. The greatest percentage of bitumen in any sample was 57.55, and the least 45.

In addition to the description of the hydraulic crusher furnished by Mr. Martin, a mention of the new Blake crusher may be of interest. See figure.

In it the old cast-iron frame is replaced by steel rods and a wooden frame, so cushioned that the fly-wheel, in case of abnormal resistance, can make a part of a revolution. The pitman, which is above the driving-axle, can be lengthened or shortened, so as to increase or diminish the stroke of the jaws, and the length of the toggles has been increased. It is probable that these changes will reduce the expense of stone-breaking by diminishing the breakage account, which, particularly with trap rock, is a serious item.

Mr. J. L. Gillespie, C. E., gives me the following as the cost of breaking 15,150 cubic yards of limestone during 1874–5–6–7, for the concrete used in the preservation of the Falls of St. Anthony. The machine, an 8-inch by 16-inch, old pattern Blake, was run by water power, for which there was no charge.

The cost and quantities were as follows, the stone being delivered at the breaker :

1874–5.....................3,452 cubic yards at.......................41.	cents.	
1875–6.....................8,284 " 81.8	"	
1876–7.... ....3,414 " ....18.	"	

One cubic yard of stone produced about two cubic yards of broken stone, the void spaces in which amounted to 50 8-10 per cent. The breakages were confined to some rubber springs, one back block, and two sets of jaws and cheeks.

The cost of breaking trap on the Palisades is given as follows, the stone being sledged, to go into the jaws readily.

Two crushers deliver 35 cubic yards of 2-inch stone per day, when working well ; 15 per cent of the time is lost by breakdowns.

COST.

1 engineman and fireman, at $2.50 .... $2.50	
2 laborers, feeding .... 1.25 .... 2.50	
2 " screening .... 1.25 .... 2.50	
	$7.50
Coal, 1 ton	3.50
Oil and waste	1.00
Breakages	5.00
	$17.00

or, say, 57 cents per cubic yard.

On Snake Island three crushers were arranged in a row, and the broken stone was carried by an endless belt to the revolving screen, whence it fell into the bins, so that no screeners were employed. The engine had one cylinder, 8 inches by 24 inches, and was running with 80 pounds of steam. The product was said to be 180 cubic yards per day when there was no breakdown.

COST.

1 engineman and fireman, at $2.50 .... $2.50	
3 laborers, feeding .... 1.25 .... 3.75	
	$6.25
2½ tons of coal, at $3.50	8.75
Oil, etc	2.00
Breakages	15.00
	$32.00

Allowing for the 15 per cent. lost by breakdowns, the cost would be about 21 cents per cubic yard.

At another place on the Hudson, two crushers, set face to face, 9-inch by 15-inch jaws, could deliver at the rate of 120 cubic yards per day, when no trouble occurred, but 100 cubic yards was a fair average.

COST.

Engineman and fireman $2.50	
3 feeders 3.75	
2 screeners 2.50	
	$8.75
1½ tons coal, at $4	6.00
Oil, etc	2.25
Repairs	10.00
	$27.00

or 27 cents per cubic yard.

NOTE.—A detailed statement of the cost, in time, of breaking stone is given in the Memoir on the Construction of a Masonry Dam, by J. J. R. Croes. Transactions of The American Society of Civil Engineers, No. CIII., Vol. VIII., page 856 (February, 1875).—[EDITOR.]

INDEX TO READING ARTICLES

ON

Streets, Highways and Paving Materials,

IN

ENGINEERING NEWS.

NOTE —This index refers only to reading articles of greater or less length on these various topics. Much further information as to prices paid in various localities will be found under the head of "News of the Week." Numerals refer to the page and Roman letters to volume.

www.ingramcontent.com/pod-product-compliance
Lightning Source LLC
Chambersburg PA
CBHW020552270326
41927CB00006B/809